Crystal Research and Technology

1

Journal of Experimental and Industrial Crystallography

Zeitschrift für experimentelle und technische Kristallographie

Volume 23
1988
Number 1 January

Akademie-Verlag · Berlin

ISSN 0232-1300 Cryst. Res. Technol., Berlin **23** (1988) 1, 1–140, K1–K28

Directions to the author

In the journal "Crystal Research and Technology – Kristall und Technik" as a rule the following categories of papers will be printed: original papers, review articles, short notes, and book reviews, preferentially in **English.** Complete information necessary for type-setting and a "Short Title" should be added.

The manuscript should not exceed 10 pages and be type-written on one side only in the format A 4 (21 × 29,7 cm) on white ink-proof paper, with a two-lined distance and a left-hand free edge of 3 cm, and submitted in 2 copies. All sheets of the paper belonging to the contribution are continuously to be numbered.

The head of the manuscript should be arranged in this manner: initials and name of the author; institution, and title of the paper. Then two abstracts should follow, one in the original language and the second in German, English, French, or Russian, respectively.

Formulae, signs and symbols, letters etc. being wanting on the type-writer, should be distinctly written with penholder. O as zero is to be underlined, the capital O is not. Greek letters should be underlined with red ink.

Paragraphs should be marked by indenting of at least 3 touches.

Concerning the marking of the manuscript should be noted:
headings should be underlined, sub-headings not, and indented by 9 touches. The titles of tables are not to be underlined. Other markings in the text will be done by the editorial staff. They mean:

light-red ∿∿∿∿∿∿∿ = small capitals
blue ∿∿∿∿∿∿∿ = in italics

light-red | perpendicular straight red line on the left edge = small printing

Tables are to be added on separate sheets of paper at the end of the contribution.

Foot notes should be continuously written on a separate sheet of paper.

Figures (Fig.) – drawings and photograäps – should be not more than 8 and submitted (2 copies) in a special envelope: photographs as white high-gloss prints (inscription on a transparent covering sheet) and drawings as distinct sketches. On the back of the submitted figures the reduction scale in tenth (9/10, 8/10, 5/10 e.g.), author's name, short title of the paper, and the number of the figures should be given.

Captions for the figures should be submitted on a separate sheet of paper.

References should be written without numbers in alphabetical order of the names at the end of the contribution. Cited names in the text stand for the references. The references should contain the following items, for example:

for books: RAST, K.: Physikalisch-chemische Rechnungen, Berlin 1958
for journals: KRAUSE, H. N., MÜLLER, K.: Cryst. Res. Technol. **16**, 70 (1981)

Add, please, the short title, about 6 keywords, and the academic grades of the authors.
The authors will be sent two galley proofs of review articles and original papers. Proof reading of short notes and book reviews will be done by the editorial staff.

For each of these papers a total of 50 reprints is available free of charge. Requests for additional maximum 200 reprints (against payment) should be given together with the submitted manuscript.

Unserem lieben, hochverehrten
Herrn Prof. Dr. Hermann Neels
zu seinem 75. Geburtstag
überreicht von seinen Kollegen
aus Druckerei und Verlag

Januar 1988

Founded 1966 as
Kristall und Technik

Crystal Research and Technology

1

VOLUME 23 · 1988 · NUMBER 1

Journal of Experimental and Industrial Crystallography

Akademie-Verlag · Berlin

Papers should be directed to the editor's office,
Scharnhorststraße 20, DDR-7030 Leipzig,
or to the Editor-in-Chief Prof. Dr. Hermann Neels,
Zeititzer Weg 25, DDR-7254 Machern,
or to one of the members of the Editorial Boad or Advisory Board.

Orders can be sent

- in the GDR:
  to the Postzeitungsvertrieb or to the Akademie-Verlag Berlin, Leipziger Straße 3 – 4, DDR-1086 Berlin;
- in the other socialist countries: to a book-shop for foreign language literature or to the competent news-distributing agency:
- in the FRG and Berlin (West): to a book-shop or to the wholesale distributing agency Kunst und Wissen, Erich Bieber OHG, Wilhelmstr. 4 – 6, D-7000 Stuttgart 1;
- in the other Western European countries: to Kunst und Wissen, Erich Bieber GmbH, General Wille-Str. 4, CH-8002 Zürich;
- in USA and Canada: to the VCH Publishers, Inc., 303 NW 12th Avenue, Deerfield Beach, FL 33442-1788, USA;
- in other countries; to the international book- and journal-selling trade, to Buchexport, Volkseigener Außenhandelsbetrieb der DDR, Postfach 160, DDR-7010 Leipzig, or to the Akademie-Verlag Berlin, Leipziger Straße 3 – 4, DDR-1086 Berlin.

Journal „Crystal Research and Technology"

Herausgeber: Im Auftrage des Akademie-Verlages Berlin herausgebeben von Prof. Dr. Hermann Neels.

Verlag: Akademie-Verlag Berlin, Leipziger Str. 3 – 4, DDR-1086 Berlin; Fernruf: 2236221 oder 2236229; Telex-Nr.: 114420; Bank: Staatsbank der DDR, Berlin, Kto.-Nr. 6836-26-20712.

Chefredakteur: Prof. Dr. Hermann Neels, Zeititzer Weg 25, DDR-7254 Machern; Fernruf: Brandis 3442. Anschrift der Redaktion: Scharnhorststr. 20, DDR-7030 Leipzig; Telefon: 310502.

Veröffentlicht unter der Lizenznummer 1526 des Presseamtes beim Vorsitzenden des Ministerrates der Deutschen Demokratischen Republik.

Gesamtherstellung: VEB Druckerei „Thomas Müntzer", DDR-5820 Bad Langensalza.

Erscheinungsweise: Das Journal „Crystal Research and Technology" erscheint monatlich. Die 12 Hefte eines Jahrganges bilden einen Band. Bezugspreis je Band 480,– DM zuzüglich Versandspesen; Preis je Heft 40,– DM. Der gültige Jahresbezugspreis für die DDR ist der Postzeitungsliste zu entnehmen.

Bestellnummer dieses Heftes: 1082/23/1.

AN (EDV) 35830
02800

Cryst. Res. Technol.	23	1988	1	3–16

Review

B. SCHUMANN, A. TEMPEL, G. KÜHN

Sektion Chemie, WB Kristallographie, Karl-Marx-Universität Leipzig

Epitaxy of $A^IB^{III}C_2^{VI}$ Semiconductors

Dedicated to Professor Hermann NEELS on the occasion of his 75th birthday

Epitaxial layers of $A^IB^{III}C_2^{VI}$ compounds were grown by the flash evaporation technique and investigated using reflection electron diffraction. Geometric-structural considerations allow the prediction of epitaxial relationships and types of epitaxy (one- or three-dimensional, multiple orientation). These properties of the epitaxial layers were confirmed experimentally. Furthermore, the epitaxial temperature ranges, non-equilibrium phases, defects, and special features of the epitaxial growth are described. Especially, the influence of lattice misfit is discussed.

Epitaktische Schichten von $A^IB^{III}C_2^{VI}$-Halbleitern wurden mittels Flash-Verdampfung dargestellt und durch Reflexionselektronenbeugung untersucht. Die mit geometrisch-strukturellen Betrachtungen vorausgesagten Verwachsungsgesetze und -typen (ein- und dreidimensionale Epitaxie, Vielfachorientierung) konnten experimentell bestätigt werden. Weiterhin werden solche Parameter wie Epitaxietemperaturbereich, Nichtgleichgewichtsphasen, Defekte und Besonderheiten des epitaktischen Wachstums beschrieben. Der Einfluß der Gitterfehlpassung ist ausführlich diskutiert.

1. Introduction

Ternary $A^IB^{III}C_2^{IV}$ compounds (A: Cu, Li, Ag; B: Al, Ga, In; C: S, Se, Te) crystallize either in the tetragonal chalcopyrite structure (space group $I\bar{4}2d$) or in the rhombic β-$NaFeO_2$ structure (space group $Pna2_1$). Especially the Cu and Ag compounds have received considerable attention for possible device applications. Thus $AgGaS_2$ and $AgGaSe_2$ have large non-linear coefficients for frequency up-conversion (second harmonic generation) and mixing as well as large electro-optical coefficients (PAMPLIN et al.). On the other hand Cu–In and Cu–Ga compounds are of technological interest because they show promise for applications as solar cells (KAZMERSKI, WAGNER). For example, using $CuInSe_2$ as an adsorber for thin film devices conversion efficiencies greater than 10% have been found (COUTTS et al.). In this connection it is important to prepare epitaxial layers, because many of the material parameters determining the efficiency of devices can be improved compared to polycrystalline films.

In previous papers we have already discussed some results of the epitaxial growth of these compounds (NEUMANN et al. 1981, SCHUMANN et al. 1986). The aim of this article is to describe the state of the art in this field more generally. An important role is played by geometrical foundations and structural investigations of $A^IB^{III}C_2^{IV}$ layers deposited on $A^{III}B^V$, Ge and CaF_2 substrates.

2. Geometric-structural relations between substrate and deposit

If one understands epitaxy as a regular and oriented growth of two crystalline substances, the substrate symmetry will have a strong influence on the symmetry of the deposited layer. This means that the symmetry properties of the epitaxial film can deviate from those of the deposit bulk material. In particular, it is even possible that the deposit can overgrow the substrate with a structure belonging to a point symmetry group which is different from that of the deposit bulk material (non-equilibrium phases, see also 4.1). Generally, from the geometric-structural point of view the following cases can occur in an epitaxial system:

(i) The symmetry of the deposited layer is exactly the same as the symmetry of the bulk material of the particular compound. In this case the deposit is grown as a single crystal. The kind of epitaxy is called *three-dimensional.*
(ii) The symmetry of the layer is different from that of the deposit bulk material, especially an ∞-fold axis is contained, the axis of texture.
The influence of the substrate consists in fixing a crystallographic direction only and the deposit crystallites are distributed around this axis randomly. This kind of epitaxy is called *one-dimensional.*
(iii) Epitaxial overgrowth is also one-dimensional yet there is not an ∞-fold axis but only an n-fold symmetry of the substrate surface.
In this case the effect of *multiple orientation* of the deposit (one-dimensional epitaxy with preferred azimutal orientation) is obtained. One must expect such a situation when the point symmetry group of the substrate offers different equivalent possibilities of epitaxial overgrowth for the deposit.

One can get a qualitative survey of the types of overgrowth in epitaxial systems by introducing the point symmetry group $S(hkl)$ of crystallographic faces $\{hkl\}$. The plane point symmetry group $S(hkl)$ is given by the following relation:

$$S(hkl) = S_{\text{cryst}} \cap \infty m|_{x_\infty // n(hkl)} .$$

Here S_{cryst} denotes the point symmetry group of the bulk material and ∞ m is the Curie group of a static cone. The intersection of the two groups is to form under the boundary condition that the symmetry axis X_∞ of the cone is parallel to the normal $\boldsymbol{n}$ of the crystallographic plane with the indices (hkl).

Now, three-dimensional epitaxy is expected in all cases where the plane point symmetry group $S_S(hkl)$ of the substrate surface is completely contained in that of the deposit plane growing parallel to the substrate surface $S_D(hkl)$. In all other cases one must expect multiple orientation:

$$S_S(hkl) \subseteq S_D(hkl) \rightarrow \text{three-dimensional epitaxy},$$

$$S_S(hkl) \nsubseteq S_D(hkl) \rightarrow \text{multiple orientation}.$$

So one can predict the effect of multiple orientation for a given epitaxial system if it is known which crystallographic deposit plane grows parallel to that substrate face forming the substrate surface.

The epitaxial relationships between tetrahedrally-coordinated semiconductors, however, can be found from the concept of corresponding crystallographic planes and rows. Tetrahedrally-coordinated semiconductors crystallize in structures belonging either to the diamond family (sphalerite, chalcopyrite) or to the lonsdaleite family (wurtzite, β-$NaFeO_2$). The two families differ only in the stacking sequence of double planes (ABCABC . . . diamond family, ABABAB . . . lonsdaleite family), which means that these families are homotectic (Megaw). Thus it is possible to find, for a given plane $\{hkl\}$ or direction $\langle uvw \rangle$ of one of the structures, the corresponding planes $\{h'k'l'\}$ or directions $\langle u'v'w' \rangle$ of all the structures of its own family or the homotectic family. For this reason one must arrange the structures in such a manner that in each case the directions of two non-collinear bonds lie parallel (see Fig. 1). Then all parallel crystallographic planes and rows in the structures are corresponding planes and rows, respectively. One can say that a *correspondence of the first kind* is given when the mesh metric formed by the atoms is onefold equal. Otherwise, when the mesh metric is manyfold equal a *correspondence of the second kind* occurs.

The Miller indices of the corresponding planes and rows can be obtained by transformation by means of appropriate matrices when the different structures are oriented relative to each other as shown in Figure 1. Table 1 gives a survey of the corresponding crystallographic planes and rows for all the structures of tetrahedrally coordinated semiconductors considered here.

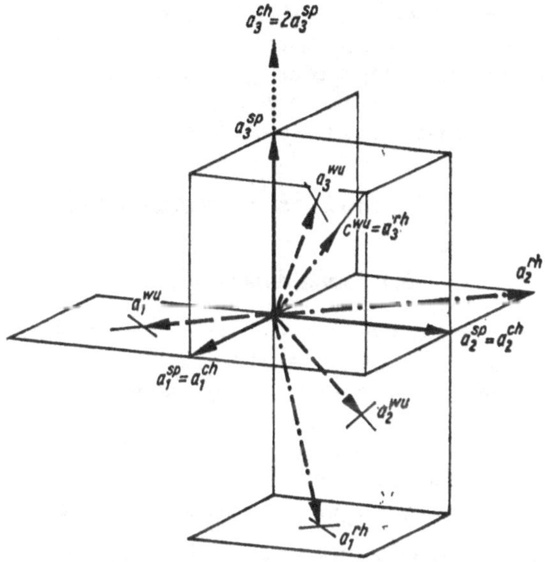

Fig. 1. Structure orientations relative to each other. sp sphalerite, ch chalcopyrite, wu wurtzite, rh rhombic

Table 1
Corresponding planes and rows in tetrahedrally coordinated semiconductors with sphalerite (sp)-, chalcopyrite (ch)-, wurtzite (wu)-, and β-$NaFeO_2$ (rh)-structure

planes				rows (directions)			
sp $\{hkl\}$	ch $\{hkl\}$	wu $\{hkil\}$	rh $\{hkl\}$	sp $\langle uvw \rangle$	ch $\langle uvw \rangle$	wu $\langle uvtw \rangle$	rh $\langle uvw \rangle$
100	100	$30\bar{3}4$	302	100	100	$40\bar{4}3$	233
	001		364		001		403
110	110	$11\bar{2}0$	010	110	110	$11\bar{2}0$	010
	102	$30\bar{3}8$	320		201	$20\bar{2}3$	210
			304				203
			368				236
111	112	0001	001	111	221	0001	001
		$30\bar{3}2$	301			$80\bar{8}3$	803
			362				463

Both for physical and for chemical reasons one can assume that in an epitaxial process tetrahedrally-coordinated semiconductors corresponding crystallographic planes and directions grow parallel. Of course, correspondences of the first kind will be preferred to those of the second kind.

As a result it is possible to predict the epitaxial relationships in an epitaxial system from the concept of corresponding lattice planes and rows — and with a knowledge of the relationships of epitaxy, one can predict whether multiple orientation will take place or not. For example, from Table 1 we find, for an epitaxial system consisting of a {100}-oriented sphalerite-type substrate and a chalcopyrite-type deposit, the epitaxial relationships

$$\{100\}_{ch} \,//\, \{100\}_{sph}\,; \quad \langle 110\rangle_{ch} \,//\, \langle 110\rangle_{sph} \text{ and}$$

$$\{001\}_{ch} \,//\, \{100\}_{sph}\,; \quad \langle 201\rangle_{ch} \,//\, \langle 110\rangle_{sph}\,.$$

The plane symmetry of the substrate surface is given by $\bar{4}3\,m \cap \infty m\,|_{X_\infty // \langle 100\rangle_S} = mm$ and the plane symmetries of the corresponding deposit planes are $\bar{4}2\,m \cap \infty m\,|_{X_\infty // \langle 001\rangle_D} = mm$ and $\bar{4}2\,m \cap \infty m\,|_{X_\infty // \langle 100\rangle_D} = 2$.

Therefore, for the first epitaxial relationship the effect of multiple orientation is to be expected. If the second is valid we have three-dimensional epitaxy.

3. Growth and investigation of the layers

3.1. Layer deposition

Some years ago, the thermal evaporation of some $A^{I}B^{III}C_2^{VI}$ compounds was investigated by STRELCHENKO et al. using a mass spctrometer combined with a Knudsen cell. These studies clearly showed that all compounds are characterized by incongruent evaporation. In the gaseous phase molecules of the types $B_2^{III}C^{VI}$ and C_2^{VI} or only C_2^{VI} could be detected. For example, the decomposition of $CuInSe_2$ follows the approximate reaction

$$2\,CuInSe_2(s) \rightarrow Cu_2Se(s) + In_2Se(g) + Se_2(g)$$

as recently found by HÖNLE and KÜHN. Therefore, it is advantageous for maintaining the stoichiometry of the compounds to use the flash-evaporation method.

The layers were grown under high vacuum conditions of about $2 \cdot 10^{-3}$ Pa (HOTTMANN; NEUMANN et al. 1981; SCHUMANN et al. 1986). Prereacted material with powder grain sizes of about 150 μm was gravity fed to a resistance-heated molybdenum or tungsten boat held at 1570 K for the most substances evaporated. The substrates were commonly polished chemomechanically and were located approximately 60 mm above the heater. In most cases the substrate temperatures were varied in steps of 25 K. Typically deposition rates and film thicknesses amounted to 0.5 nm s^{-1} and 100 nm.

Some experiments were carried out by using the single-source evaporation technique (NEUMANN et al. 1980).

3.2. Structural investigations

The layers were investigated by reflection high energy electron diffraction (RHEED) measurements at accelerating voltages of 50 or 65 KV.

Because of the microscopical roughness of the flash-evaporated epitaxial films the incident electron beam will penetrate the tops of small hillocks forming the layer surface and thus give a diffraction pattern which is a transmission diagram rather than a reflection diagram.

Nevertheless, such diagrams obtained under vacuum conditions of about 10^{-3} Pa consist of a great number of reflection spots and permit one to obtain much information about the epitaxial layers. Of course, one can determine the epitaxial relationships between the deposit and the substrate from such diffraction patterns, but besides this the diagrams contain information about the defect structure of the layers (e.g. other crystalline phases, twin formation, stacking faults, grain boundaries, and the degree of epitaxy).

In this connection it seems to be necessary to make some remarks about the structure determination of $A^{I}B^{III}C_2^{VI}$ epitaxial layers by means of the RHEED method in more detail.

The structures of chalcopyrite and β-$NaFeO_2$ differ from those of sphalerite and wurtzite, respectively, in an alternate occupation of the metal atom positions by different atoms giving a lower symmetry. E.g., the point symmetry group of the chalcopyrite structure is a subgroup of the sphalerite point symmetry group. From this it follows that RHEED patterns of structures within the same family contain all the reflections of the structures showing a higher symmetry in this family. Especially, the aristotype of the structure family gives the smallest number of reflection spots within the family. Therefore, it is possible to distinguish by RHEED a lower symmetrical phase besides parts with higher symmetry in an epitaxial layer because additional spots will occur but it is impossible to find parts of a higher symmetrical phase besides those with a lower symmetry.

In the case of $CuGaSe_2$ (SCHUMANN et al. 1978b), however, the additional spots of the chalcopyrite phase have a very small intensity as compared to those of the sphalerite structure because the atomic form factors of Cu and Ga are nearly the same and so it is very difficult to distinguish between the sphalerite and the chalcopyrite structures by RHEED investigations. But in this case the deviation of the c/a-ratio from the ideal value 2.0 is sufficient to make a reflection splitting when the effect of multiple orientation has taken place in the epitaxial $CuGaSe_2$ layers.

Further from such reflection splitting in RHEED diagrams the c/a-ration can be determined with an accuracy of about 10^{-4} without a knowledge of the lattice constants a and c (TEMPEL, SCHUMANN 1978, 1979a).

The numerical values of the lattice constants of $A^{I}B^{III}C_2^{VI}$ epitaxial layers cane be determined by RHEED measurements using the substrate as calibrating material. If RHEED diagrams contain reflections both of the substrate and the deposit the accuracy of this method is about 10^{-2} ... 10^{-3} (TEMPEL, SCHUMANN 1979b).

4. Results and discussion

4.1. Survey of the epitaxial systems investigated

In Table 2 the epitaxial systems studied up to now are listed and the epitaxial relationships found experimentally as well as the temperature range ΔT_E for epitaxial growth are given. In addition to the compiled data mixed crystal epitaxial layers of $CuInSe_2$–$CuGaSe_2$ (SCHUMANN et al. 1980), $CuInTe_2$–$LiInTe_2$ (SCHUMANN et al. 1987) and $CuInSe_2$–$LiInSe_2$ (MITARAY et al.) were grown. Furthermore, we found epitaxial growth of $A^{I}B^{III}C_2^{VI}$ compounds on NaCl substrates.

In some cases in the deposition on {111}-oriented substrates at high deposition temperatures the sphalerite phase of the deposit was observed (Tab. 3). However,

Table 2
Epitaxial systems investigated

substrates		deposite	structure	epitaxial relationships (experimental)				type of epitaxy	ΔT_E/K	references
compound	orientation {hkl}	compound		$\{hkl\}_S$ //	$\{hkl\}_D$	$\langle uvw\rangle_S$ //	$\langle uvw\rangle_D$			
GaAs	111	$CuInSe_2$	ch	111	112	110	110	a	720 – 870	SCHUMANN et al. 1978a SCHUMANN et al. 1981
		$CuInS_2$	ch					a	645 – 870	NEUMANN et al. 1979
		$CuInTe_2$	ch					a	620 – 790	NEUMANN et al. 1980
		$CuGaSe_2$	ch					a	835 – 870	SCHUMANN et al. 1978b
		$CuGaTe_2$	ch					a	695 – 840	SCHUMANN et al. 1979
		$AgGaSe_2$	ch					a	845 – 920	SCHUMANN et al. 1985 TEMPEL et al. 1985
InAs	111	$CuInTe_2$	ch					a	≈ 720	SCHUMANN et al. 1987
		$LiInTe_2$	ch					a	≈ 645	SCHUMANN et al. 1987
GaAs	111	$LiInSe_2$	rh	111	001	110	010	a	620 – 870	MITARAY TEMPEL et al. 1983
GaP	111	$LiInSe_2$	rh					a	645 – 845	MITARAY
GaAs	100	$CuInSe_2$	ch	100	001	110	110	c	720 – 920	TEMPEL et al. 1981
					100		021	a		
		$CuGaSe_2$	ch	100	001	110	110	c	870 – 895	SCHUMANN et al. 1983
GaAs	100	$AgGaSe_2$	ch	100	100	110	021	a	845 – 895	SCHUMANN et al. 1985 TEMPEL, SCHUMANN
GaP	100	$CuGaSe_2$	ch					a	795 – 895	SCHUMANN et al. 1983
InAs		$LiInTe_2$	ch					a	645	SCHUMANN et al. 1987
GaAs		$LiInSe_2$	ch					a	620 – 870	MITARAY TEMPEL et al. 1986
		$LiInSe_2$	rh	100	302	110	203	a		
					364		236	a		
GaP		$LiInSe_2$	rh	100	302	110	203	a		
					364		236			
GaAs	110	$CuInSe_2$	ch	110	012	110	021	a	720 – 920	TEMPEL et al. 1981
					110		110	c		
		$AgGaSe_2$	ch	110	012	110	021	a	845 – 895	SCHUMANN et al. 1985
				111	112	110	021			TEMPEL, SCHUMANN
		$CuGaSe_2$	ch	110	012	110	021	a	845 – 895	SCHUMANN unpubl.
		$LiInSe_2$	ch	110	110	110	110	c	670 – 870	MITARAY
		$LiInSe_2$	rh	110	304	110	010	a		TEMPEL et al. 1986
					368		210	a		
Ge	111	$CuInSe_2$	ch	111	112			b	720 – 845	SCHUMANN et al. 1980a
		$LiInSe_2$	rh	111	001			b		MITARAY
CaF_2	111	$CuInSe_2$	ch	111	112	110	110	a	775 – 800	SCHUMANN, KÜHN
		$CuInS_2$	ch					a	825 – 850	
		$CuInTe_2$	ch					a	750 – 825	
		$CuGaSe_2$	ch					a	850 – 900	
		$CuGaS_2$	ch					a	850 – 900	
		$LiInSe_2$	rh	111	001	110	010	a	670 – 820	MITARAY

ΔT_E: substrate temperature range for epitaxial growth
type of epitaxy: a multiple orientation, b one-dimensional (texture), c three-dimensional
structures: sp sphalerite, ch chalcopyrite, c β-$NaFeO_2$

Table 3
Epitaxial $A^IB^{III}C_2^{VI}$ layers with sphalerite structure (ΔT_E – substrate temperature range (K), T_u – transformation temperature (K), chalcopyrite-sphalerite in bulk material, MÖLLER et al.)

deposit	substrate	ΔT_E/K	T_u/K	reference
$CuInS_2$	GaAs {111} A	870	1270	NEUMANN et al. 1979
$CuInSe_2$	GaAs {111} A	870	1083	SCHUMANN et al. 1978a
	GaAs {111} B	695–845		SCHUMANN et al. 1981
	Ge {111}	795–845		SCHUMANN et al. 1980a
$CuGaSe_2$	GaAs {111} A	860–870	1320	SCHUMANN et al. 1978b
$CuIn_{0.7}Ga_{0.3}Se_2$	GaAs {111} A	850	–	SCHUMANN et al. 1980b

these temperatures were distinctly below the temperatures of the phase transitions in the bulk material. We suppose that the re-evaporation of B^{III}-C^{VI} species from the {112} deposit surface is higher than for the other surfaces considered here in agreement with their smaller density of dangling bonds (HARVEY, GATOS). In the case of $CuInSe_2$ a shift in the copper-rich region of the phase diagram was detected and thus the occurrence of the sphalerite phase in thin epitaxial films was explained (KÜHN, SCHUMANN).

Another type of metastable phases was found in the epitaxy of $LiInSe_2$ on {111}-oriented GaAs substrates (TEMPEL et al. 1983). $LiInSe_2$ crystallizes in the chalcopyrite structure instead of the thermodynamically stable β-$NaFeO_2$ structure. Between the two phases epitaxial intergrowth was observed with the epitaxial relationships

$$\{112\}_{ch} // \{001\}_{rh} ; \quad \langle 110\rangle_{ch} // \langle 010\rangle_{rh} .$$

The presence of the sphalerite phase cannot be excluded here.

4.2. The role of lattice misfit

Lattice misfit plays an important role when epitaxy is considered as a geometric-structural problem. It is well known that epitaxial systems are able in general to compensate misfit values up to about 15%. This limit must be understood as an approximate value which is dependent on many parameters.

For epitaxial systems of tetrahedrally coordinated semiconductors the following definitions of lattice misfit appear meaningful:

(i) average misfit $\bar{\beta}$

$$\bar{\beta} = \frac{\bar{a}_D - \bar{a}_S}{\bar{a}_S}$$

$\bar{a}_S$ average lattice parameter of the substrate
$\bar{a}_D$ average lattice parameter of the deposit
$\bar{a} = (V)^{1/3}$ for cubic unit cells (sphalerite or diamond structure)
$\bar{a} = (1/2\ V)^{1/3}$ for tetragonal or rhombic unit cells (chalcopyrite or β-$NaFeO_2$ structure)
V volume of the unit cell.

The $\bar{\beta}$ values are suitable to obtain a first idea of the size of the lattice misfit of a given substrate-deposit combination.

(ii) direction misfit β_{uvw}

$$\beta_{uvw} = \frac{r_D - r_S}{r_S}$$

r_D atomic distance of the deposit in the direction $\langle uvw \rangle_D$

r_S atomic distance of the substrate in the direction $\langle uvw \rangle_S$

β_{uvw} gives the misfit between corresponding directions of the substrate and the deposit.

(iii) planar misfit β_{hkl}

$$\beta_{hkl} = \frac{|(r_{1D} - r_{1S})\, r_{2S}| + |(r_{2D} - r_{2S})\, r_{1S}|}{r_{1S} \cdot r_{2S}} = \beta_{uvw}^1 + \beta_{uvw}^2$$

r length of the lattice vectors r_1 and r_2 (atomic distances), forming the corresponding unit meshes of deposit (D) and substrate (S) in the intergrowth plane with the corresponding direction misfits β_{uvw}^1 and β_{uvw}^2.

Concerning the epitaxy of $A^IB^{III}C_2^{VI}$ layers on $A^{III}B^V$ and Ge substrates there is for all these systems a complete epitaxial growth (no polycrystalline parts detectable with RHEED) listed in Table 2. The lattice parameters of the substances and the values of average misfit $\bar{\beta}$ of the deposits with GaAs are given in Table 4.

The investigation of the epitaxial behaviour of the mixed crystal system Cu_xLi_{1-x}–$InTe_2$ (SCHUMANN et al. 1987) showed that there is complete epitaxial growth on The investigation of the epitaxial behaviour of the mixed crystal system— GaAs substrates only for $x \geqq 0.5$ ($\bar{\beta} \leqq 10.9\%$). The lithium-rich films always contained polycrystalline parts. However, on InAs substrates complete epitaxy was found ($\bar{\beta}$: 2.2 ... 5.0%). Incomplete epitaxial growth was observed also for Cu_xLi_{1-x}–$InTe_2$ films on GaP substrates ($\bar{\beta} \geqq 13.4\%$). Therefore, these results allow the conclusion that the given growth conditions require an average lattice misfit smaller than about 12% for complete epitaxy of $A^IB^{III}C_2^{VI}$ compounds on $A^{III}B^V$ substrates, which crystallize all in related tetrahedrally coordinated structures.

Table 4
Lattice parameters a, c, $\bar{a}$ (nm) and average misfit $\bar{\beta}$ (%) between $A^IB^{III}C_2^{VI}$ layers and GaAs substrates

compound	a	c	$\bar{a}$	$\bar{\beta}$
GaAs	0.56533			
GaP	0.54504			
InAs	0.60548			
CaF_2	0.5464			
NaCl	0.56401			
$CuInS_2$	0.552	1.108	0.5527	−2.2
$CuInSe_2$	0.5782	1.162	0.5791	2.5
$CuGaSe_2$	0.5612	1.1032	0.5580	−1.3
$AgGaSe_2$	0.5895	1.096	0.5801	2.6
$CuGaTe_2$	0.600	1.193	0.5988	5.9
$CuInTe_2$	0.6188	1.2406	0.6193	9.6
$Cu_{0.5}Li_{0.5}ImTe_2$	0.6277	1.2495	0.6267	10.9
$LiInTe_2$	0.6418	1.2498	0.6362	12.5

Table 5
Epitaxial relationships and planar lattice misfit β_{hkl} (%) for $A^IB^{III}C_2^{VI}$ deposits with chalcopyrite structure on {100}- and {100}-oriented $A^{III}B^V$ substrates

compounds	substrate orientation {hkl}	deposit orientation {hkl}	β_{hkl}	epitaxial relationship (experimental)
$CuInSe_2$/GaAs	100	100	5.05	(−)
	100	001	4.55	(+)
$CuInSe_2$/GaAs	110	110	5.05	(−)
	110	012	4.80	(+)
$AgGaSe_2$/GaAs	100	100	9.46	+
	100	001	11.73	−
$AgGaSe_2$/GaAs	110	110	9.46	−
	110	012	7.12	+
$CuGaSe_2$/GaAs	100	100	3.16	−
	100	001	1.46	+
$CuGaSe_2$/GaP	100	100	4.17	+
	100	001	5.92	−
$CuGaSe_2$/GaAs	110	110	3.16	−
	110	012	2.31	+
$LiInTe_2$/InAs	100	100	9.10	+
	100	001	11.87	−

+ realized, − not realized, (+) preferred, (−) not preferred

In all the $A^IB^{III}C_2^{VI}$ layers deposited on CaF_2 substrates there were polycrystalline besides epitaxial parts in the films independently of the average lattice misfit $\bar{\beta}$, which was in the range from 1.2 to 12.8% (SCHUMANN, KÜHN). Probably, this is caused by the strong ionic interaction at the interface and by the large differences in the thermal expansion coefficients of the two materials.

In the deposition of $A^IB^{III}C_2^{VI}$ compounds on {100}- and {110}-oriented substrates there are always different orientation possibilities for the deposit (see Ch. 2). In these cases one would expect that the epitaxial relationships having the smallest planar misfit β_{hkl} are preferred. This influence of misfit was observed in all the epitaxial systems investigated. The results are summarized in Table 5. Different misfits occur if the c/a-ratio of a deposit with chalcopyrite structure differs from 2.0. The differences in the planar lattice misfits increase with increasing deviation of the c/a-ratio from the ideal value 2.0. However, even in the deposition of $CuInSe_2$ ($c/a = 2.01$) on {100}- and {110}-oriented GaAs substrates this effect is evident (TEMPEL et al. 1981). Indeed, both orientation possibilities of the deposit still appear but that with the smaller misfit is preferred. For $CuGaSe_2$ ($c/a = 1.96$) (SCHUMANN et al. 1983), $LiInTe_2$ ($c/a = 1.95$) (SCHUMANN et al. 1987), and $AgGaSe_2$ ($c/a = 1.82$) (SCHUMANN et al. 1985) only the optimum variant was observed. Using different substrates for a deposit compound one can obtain different epitaxial relationships, for example:

$$\{001\}_{CuGaSe_2} // \{100\}_{GaAs} ; \quad \langle 110 \rangle_{CuGaSe_2} // \langle 110 \rangle_{GaAs} ,$$

$$\{100\}_{CuGaSe_2} // \{100\}_{GaP} ; \quad \langle 021 \rangle_{CuGaSe_2} // \langle 110 \rangle_{GaP} .$$

4.3. One-dimensional epitaxy

Most of the epitaxial layers investigated showed multiple orientation or partly tree-dimensional epitaxial growth according to the considerations of Chapter 2. However,

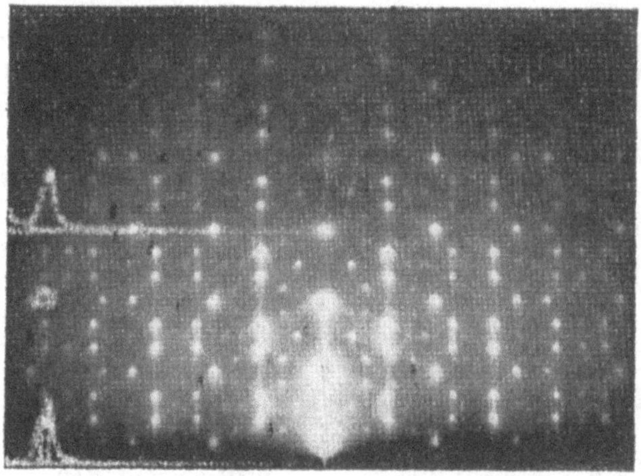

Fig. 2. RHEED diagram of a $CuInSe_2$ thin film on {111}-oriented germanium substratesl $T_S = 720$ K. The film has chalcopyrite structure and the type of epitaxy is one-dimensiona (texture)

using {111}-oriented germanium as substrate, the epitaxy of $CuInSe_2$ (SCHUMANN et al. 1980a) and $LiInSe_2$ (MITARAY) is not characterized by multiple orientation, as would be expected, but is one-dimensional (texture).

The electron diffraction diagrams (see e.g. Fig. 2) are completely invariable against azimuthal rotation of the samples. The epitaxial relationships are:

$$\{111\}_{Ge} \,//\, \{112\}_{CuInSe_2} \quad \text{and}$$

$$\{111\}_{Ge} \,//\, \{001\}_{LiInSe_2} \,.$$

The deposit crystallites are exactly aligned relative to the substrate surface normal, which is the texture axis, but parallel to the substrate surface they are distributed statistically (see also Fig. 2). It should be noted that in the epitaxy of $CuInSe_2$ on Ge this is valid for both the chalcopyrite and the sphalerite phase of the deposit at $T_S \lessapprox 795$ K.

Probably, this texture is caused by the great covalent part in the chemical bond of Ge compared with that of the $A^{III}B^V$ substrates, especially GaAs. Epitaxial deposition of $A^IB^{III}C_2^{VI}$ compounds on Si substrates does not occur.

4.4. Epitaxy of $AgGaSe_2$ on GaAs substrates

Of all the known $A^IB^{III}C_2^{VI}$ compounds crystallizing in the chalcopyrite structure $AgGaSe_2$ shows one of the greatest deviations ε from the ideal c/a-ratio which is 2.0 ($\varepsilon = -0.18$). The amount of this deviation gives rise to special features in the RHEED diagrams which allow one to obtain some essential information about the epitaxial system. Particularly, it is possible to find the accurate epitaxial relationship (exact direction coincidence) and to determine the atomic structure of the interface between the substrate and the deposit (SCHUMANN et al. 1985; TEMPEL).

When epitaxial growth takes place between $A^IB^{III}C_2^{VI}$ chalcopyrite-type deposits and sphalerite-type substrates with {111} orientation there are parallel {112} deposit and {111} substrate planes. The possible coincidences of crystallographic directions are $\langle 110\rangle_S \,//\, \langle 110\rangle_D$ and $\langle 110\rangle_S \,//\, \langle 021\rangle_D$. For reasons of symmetry and misfit one must expect the first relation to be preferred.

An analysis of diffraction patterns from $AgGaSe_2$ epitaxial layers compared to the substrate Kikuchi lines obtained in exact $\langle 211\rangle_S$ azimuts gives the following results (TEMPEL et al. 1985):

– The exact epitaxial relationship is given by

$$\{111\}_S \,//\, \{112\}_D\ ;\quad \langle 110\rangle_S \,//\, \langle 110\rangle_D\,.$$

Probably, this fact is generalizable for the epitaxial growth of $A^IB^{III}C_2^{VI}$ semiconductors on sphalerite-type substrates.

– Twin formation is detectable also in $\langle 211\rangle_S$ azimuths.

– The c/a-ratio was found to be 1.815 ± 0.005.

The epitaxial growth of $AgGaSe_2$ layers on $\{100\}$-oriented GaAs substrates is characterized by the relationships

$$\{100\}_S \,//\, \{100\}_D\ ;\quad \langle 110\rangle_S \,//\, \langle 021\rangle_D\,.$$

In all the epitaxial layers of $AgGaSe_2$ on $\{100\}$-oriented GaAs the coincidence of crystallographic $\langle 110\rangle_S$ and $\langle 021\rangle_D$ directions was found, which are the most densely packed lattice rows and are PBC-vectors in their structures. However, such coincidences occur only in *one* of the two $\langle 110\rangle_S$ directions what is shown schematically in Figure 3 (see also TEMPEL, SCHUMANN 1986). Because the angle between the two $\langle 021\rangle$ deposit directions is different from 90 degrees this situation can be proved by RHEED.

From this and by means of geometric-structural considerations one can conclude that the interface in the epitaxial system $AgGaSe_2$/GaAs consists either of a Ga and a Se layer or of a As and a Ag–Ga layer. In connection with etch experiments it is possible to distinguish the two variants. The result is that the uppermost layer of the GaAs substrate is a Ga layer and the deposit begins with a layer of Se atoms (TEMPEL).

Further in the case of $\{110\}$-oriented GaAs substrates (TEMPEL, SCHUMANN 1986) the epitaxial relationships were found to be dependent on the growth temperature. At growth temperatures of 870 K there are no low indexed deposit lattice planes which are parallel to the $\{110\}$ substrate surface. Here the most densely packed lattice planes $\{111\}_S$ and $\{112\}_D$ were found to be parallel and so the exact epitaxial

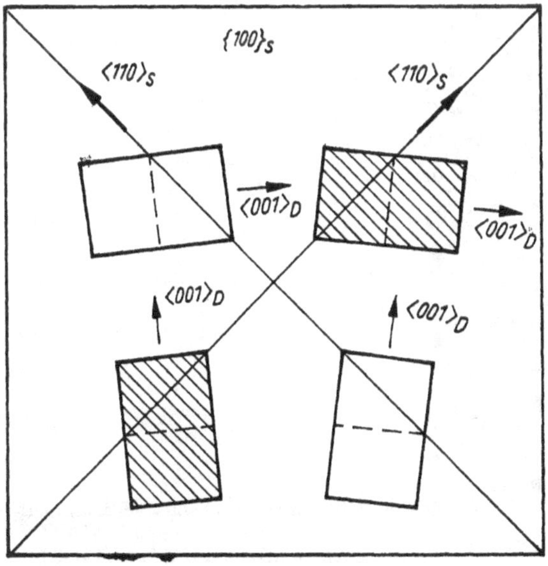

Fig. 3. Schematic representation of the possibilities for orientation of $AgGaSe_2$ crystallites on $\{100\}$-oriented GaAs substrates

relationship must be described as

$$\{111\}_S \,//\, \{112\}_D\,; \quad \langle 110\rangle_S \,//\, \langle 110\rangle_D\,, \quad \langle 021\rangle_D\,.$$

At growth temperatures of 895 K the relationships are given by

$$\{110\}_S \,//\, \{021\}_D\,; \quad \langle 110\rangle_S \,//\, \langle 021\rangle_D$$

because of the smaller planar misfit in comparison with $\{110\}_D$.

From the special geometric-structural features of the epitaxy of $AgGaSe_2$ with the high deviation of the c/a-ratio from the value 2.0 on GaAs substrates one can conclude that these effects are valid for the epitaxial growth of all $A^IB^{III}C_2^{VI}$ deposits on $A^{III}B^V$ substrates and especially the $\langle 111\rangle$ A directions of the substrate (directions from a Ga atom to the next neighbouring As atom) play an important role in this process.

4.5. Defects

Because of both the effect of multiple orientation and the realization of different epitaxial relationships antiphase domain boundaries and grain boundaries ($c/a \neq 2$) will exist in the epitaxial layers. In both cases the defects are planar defects limiting some physical properties of the epitaxial films.

In Figure 4 a schematic representation for antiphase domains (orientation and translation variants) is shown, which can be expected for the epitaxial growth of chalcopyrite-type semiconductors on $A^{III}B^V$ substrates. Moreover, in the case of $c/a \neq 2$ for the deposits there are angle inclinations between the domains causing the antiphase domain boundaries to be also small angle boundaries. An example of this case is the epitaxy of $CuInSe_2$ on $\{100\}$-oriented GaAs. The RHEED pattern of such films shows crystallites with different orientations according to the relationships given in Table 5. These crystallites include angles slightly different from 90 degrees because of $c/a = 2.01 \neq 2$. Therefore, the epitaxial layers are characterized by a very high dislocation density, which was proved by TEM investigations (TEMPEL, SCHUMANN unpublished results).

Fig. 4. Schematic representation of antiphase domains in chalcopyrite-type $A^IB^{III}C_2^{VI}$ layers ($c/a = 2.0$) on $A^{III}B^V$ substrates

Further, twins and stacking faults can be detected by means of RHEED measurements. Both kinds of defects depend on the substrate temperature in a manner which is not clear, but there is a tendency for these defects to decrease and then to increase as the growth temperatures within the epitaxial range increase.

If twin formation takes place simple twinning dominate especially when the substrate normal is parallel to a ⟨221⟩ deposit direction. However, in many cases there are also multiple twins within the layers.

The formation of stacking faults was observed in the directions ⟨221⟩ (chalcopyrite structure) and ⟨001⟩ (β-$NaFeO_2$ structure). Such stacking faults are connected with the occurrence of homotectic phases because each stacking faults represents a small layer of the corresponding homotectic structure.

Moreover, in many cases superlattice structures exist in the epitaxial films caused probably by an ordered distribution of vacancies, e.g. in the systems $CuInSe_2$/GaAs (100) (TEMPEL et al. 1981), $CuGaSe_2$/GaAs (100) (SCHUMANN et al. 1983) and $AgGaSe_2$/GaAs (111) (SCHUMANN et al. 1985).

References

COUTS, T. L., KAZMERSKI, L. L., WAGNER, S. (eds.): Solar Cells, vol. 16, Lausanne 1986 (Special Issue on $CuInSe_2$)

HARVEY, W. N., GATOS, H. C.: J. Electrochem. Soc. **105**, 654 (1958)

HÖNLE, W., KÜHN, G.: J. Thermal Analysis **31**, 589 (1986)

HOTTMANN, H.: Dissertation A, Leipzig 1974

KAZMERSKI, L. L., WAGNER, S.: in: Current Topics in Photovoltaics (T. L. COUTTS, J. D. MEAKIN, eds.), London 1985

KÜHN, G., SCHUMANN, B.: Cryst. Res. Technol., to be publ.

MEGAW, H. D.: Crystal Structures: A Working Approach, Philadelphia 1973

MITARAY, S.: Dissertation A, Leipzig 1984

MITARAY, S., KÜHN, G., SCHUMANN, B., TEMPEL, A., HÖRIG, W., NEUMANN, H.: Thin Solid Films **135**, 251 (1986)

MÖLLER, W., KÜHN, G., ENGEL, W.: Thermochimica Acta **78**, 129 (1984)

NEUMANN, H., KÜHN, G., SCHUMANN, B.: Progr. Cryst. Growth Charact., **3**, 57 (1981)

NEUMANN, H., NOWAK, E., SCHUMANN, B., TEMPEL, A., KÜHN, G.: Kristall u. Technik **15**, 61 (1980)

NEUMANN, H., SCHUMANN, B., PETERS, D., TEMPEL, A., KÜHN, G.: Kristall u. Technik **14**, 379 (1979)

PAMPLIN, B. R., JOSHI, N. V., SCHWAB, C. (eds.): Progr. Cryst. Growth Charact., vol. 10, Oxford 1985 (Proc. Sixth Int. Conf. on Ternary and Multinary Compounds)

SCHUMANN, B., GEORGI, C., TEMPEL, A., KÜHN, G., NGUYEN VAN NAM, NEUMANN, H., HÖRIG, W.: Thin Solid Films **52**, 45 (1978) a

SCHUMANN, B., KÜHN, G.: Cryst. Res. Technol. **19**, 1079 (1984)

SCHUMANN, B.,NEUMANN, H., NOWAK, E., KÜHN, G.: Cryst. Res. Technol. **16**, 675 (1981)

SCHUMANN, B., NEUMANN, H., TEMPEL, A., KÜHN, G., NOWAK, E.: Kristall u. Technik **15**, 71 (1980) a

SCHUMANN, B., NOLZE, G., KÜHN, G.: Thin Solid Films, to be publ.

SCHUMANN, B., TEMPEL, A., GEORGI, C., KÜHN, G.: Thin Solid Films **70**, 319 (1980) b

SCHUMANN, B., TEMPEL, A., KÜHN, G.: Cryst. Res. Technol. **18**, 71 (1983)

SCHUMANN, B., TEMPEL, A., KÜHN, G.: Solar Cells **16**, 43 (1986)

SCHUMANN, B., TEMPEL, A., KÜHN, G., NEUMANN, H.: Thin Solid Films **130**, 113 (1985)

SCHUMANN, B., TEMPEL, A., KÜHN, G., NEUMANN, H., NGUYEN VAN NAM, HÄNSEL, T.: Kristall u. Technol **13**, 1285 (1978) b

SCHUMANN, B., TEMPEL, A., KÜHN, G., NEUMANN, H., PETERS, D., HÖRIG, W.: Kristall u. Technik **14**, 665 (1979)

STRELCHENKO, S. S., BONDAR, S. A., MOLODYK, A. D., BALANEVSKAYA, A. E., BERGER, L. I.: Zh. Fiz. Khim. **41**, 3118 (1967)
STRELCHENKO, S. S., BONDAR, S. A., MOLODYK, A. D., BERGER, L. I., BALANEVSKAYA, A. E.: Isv. Akad. Mauk, SSSR. Neorg. Nater. **5**, 593 (1969)
TEMPEL, A.: Cryst. Res. Technol. **20**, K84 (1985)
TEMPEL, A.: Dissertation B, Leipzig 1986
TEMPEL, A., SCHUMANN, B.: Kristall u. Technik **13**, 389 (1978)
TEMPEL, A., SCHUMANN, B.: Thin Solid Films **59**, 99 (1979) a
TEMPEL, A., SCHUMANN, B.: Kristall u. Technik **14**, 571 (1979) b
TEMPEL, A., SCHUMANN, B.: Cryst. Res. Technol. **21**, 311 (1986)
TEMPEL, A., SCHUMANN, B., KOLB, K., KÜHN, G.: J. Cryst. Growth **54**, 534 (1981)
TEMPEL, A., SCHUMANN, B., KÜHN, G.: Cryst. Res. Technol. **20**, 1571 (1985)
TEMPEL, A., SCHUMANN, B., MITARAY, S., KÜHN, G.: Thin Solid Films **101**, 339 (1983)
TEMPEL, A., SCHUMANN, B., MITARAY, S., KÜHN, G.: Cryst. Res. Technol. **21**, 1429 (1986)

(Received April 21, 1987)

Authors addresses:

Prof. Dr. G. KÜHN, Dr. B. SCHUMANN
Karl-Marx-Universität, Sektion Chemie, WB Kristallographie
Talstraße 35
Leipzig
DDR-7010

Doz. Dr. A. TEMPEL
Technische Universität Dresden, Sektion Physik, WB Oberflächen- und Elektronenphysik
Mommsenstraße 13
Dresden
DDR-8027

Cryst. Res. Technol.	23	1988	1	17–24

G. Pfaff, A. Feltz, U. Umpfenbach

Sektion Chemie der Friedrich-Schiller-Universität Jena, DDR

Solid-State Reactivity and Mechanisms in Oxide Systems (IV)

On the Sintering Behaviour of Haematite

Dedicated to Professor Hermann Neels on the occasion of his 75th birthday

The present contribution reports about the sintering behaviour of three α-Fe_2O_3 powders which were synthesized in various ways. The different results of the sintering experiments are the consequence of granulometric properties of the samples and of their content of impurities. Possible sintering mechanisms are discussed.

Die vorliegende Arbeit berichtet über das Sinterverhalten von drei auf unterschiedlichem Wege hergestellten α-Fe_2O_3-Pulvern. Die voneinander abweichenden Ergebnisse der Sinterexperimente sind sowohl auf granulometrische Eigenschaften der Proben als auch auf deren Gehalt an Verunreinigungen zurückzuführen. Sie werden hinsichtlich möglicher Sintermechanismen diskutiert.

1. Introduction

Previous investigations have shown that α-Fe_2O_3 powder prepared in different ways is accurately characterized by the rates of reaction with small parts of ZnO (Feltz, Martin a, b) as well as in equimolar mixtures (Feltz, Martin c). Starting from a mixture of ZnO, Mn_2O_3 and Fe_2O_3 the reactivity of the latter seems to determine the properties of soft magnetic Zn, Mn-ferrites to a large degree depending on the grain size structure of the ceramics (König). It is the aim of this paper to study the reaction of α-Fe_2O_3 with itself in dependence on origin and impurity content by measuring the rate of shrinkage and grain growth as a result of sintering at different temperatures.

Coble verified several different models for sintering of α-Fe_2O_3. He found that the apparent diffusion coefficients and temperature dependence calculated from the shrinkage of pressed compacts of Fe_2O_3 agree with measured diffusion coefficients for the diffusion of Fe in Fe_2O_3. Igarashi et al. reported on rapid initial densification of impure powders assuming that the closure of very small pores present in the compacts should be responsible for the shrinkage behaviour. On the other hand, compacts made of pure α-Fe_2O_3 powder increase their density slowly with time. De Wit and Geus confirmed these results suggesting a reduction-oxidation process at the surface of the grains. In the impure material the formation of low melting eutectic phases has also been discussed.

2. Experimental

Three different kinds of α-Fe_2O_3 were applied. α-Fe_2O_3(H) prepared by hydrothermal formation from $Fe(NO_3)_3$ solution as described by Voigt and Göbler shows the lowest impurity content.

$$2\,Fe(NO_3)_3 + H_2O \xrightarrow{200\,°C} Fe_2O_3 + 6\,HNO_3\,.$$

This type consists of relatively large spherical agglomerated grains of different size. Another type of α-Fe_2O_3 powder was obtained by thermal treatment of goethite precipitated from a solution of $FeSO_4 \cdot 7\,H_2O$ under the action of oxygen and stirring at a preserved acidity (PFAFF et al.).

$$4\,FeSO_4 + 6\,H_2O + O_2 \rightarrow 4\,FeO(OH) + 8\,H^+ + 4\,SO_4^{2-}\,.$$

Annealing of α-FeO(OH) at 450 °C (2 h) followed by heating at 800 °C (2 h) yields α-Fe_2O_3(G) powder of ellipsoid grains with relatively small size distribution. Commercial α-Fe_2O_3 from BAYER (WF ztype) was included for comparison.

Some morphologic powder data and the impurity concentration of several elements determined by atomic absorption spectroscopy are summarized in Table 1. There is an important difference between α-Fe_2O_3(H) on the one side and α-Fe_2O_3 (WF) and α-Fe_2O_3(G) on the other side (Fig. 1). The latter seems to be a little more agglomerated than α-Fe_2O_3 (WF). The former consists of spherolites of relatively large size showing the lowest content of impurities. In order to study the influence of the impurities on the sintering behaviour α-Fe_2O_3(H) powder samples were doped evaporating aqueous suspensions which contained small amounts of silicic acid formed from an addition of $Si(OC_2H_5)_4$ and/or of $Mn(NO_3)_2 \cdot 4\,H_2O$, $NaNO_3$ and/or of $Al(NO_3)_3 \cdot 9\,H_2O$ or the combination of the four elements:

(a) α-Fe_2O_3(H) + 60 ppm Na,
(b) α-Fe_2O_3(H) + 120 ppm Al,
(c) α-Fe_2O_3(H) + 60 ppm Na + 120 ppm Al,
(d) α-Fe_2O_3(H) + 250 ppm Si,
(e) α-Fe_2O_3(H) + 250 ppm Si + 1000 ppm Mn,
(f) α-Fe_2O_3(H) + 250 ppm Si + 1000 ppm Mn + 60 ppm Na + 120 ppm Al.

The residue was ball milled and the powder compacted by a pressure of $3 \cdot 10^8\ N \cdot m^{-2}$ at room temperature. The tablets (10 mm in diameter with an altitude of 5 mm) attained

Table 1
Properties of the α-Fe_2O_3 types WF, G, and H

Property	α-Fe_2O_3(WF)	α-Fe_2O_3(G)	α-Fe_2O_3(H)
Specific surface area by BET [m^2/g]	4.0	2.0	1.0
$\bar{d}_{BET}$ [μm]	0.30	0.60	1.15
$d_{50\%}$ value from sedimentation analysis [μm]	0.54	2.00	7.10
d-values from ELMI [μm]	0.2–0.6	0.2–0.6	0.2–7.0
Impurities [ppm]			
Si	200	200	75
Mg	10	3	—
Al	120	100	—
Cu	25	8	2
Mn	1000	1000	—
Cr	10	—	—
Ni	100	200	50
Pb	5	—	8
Ca	2	—	5
Na	100	100	50

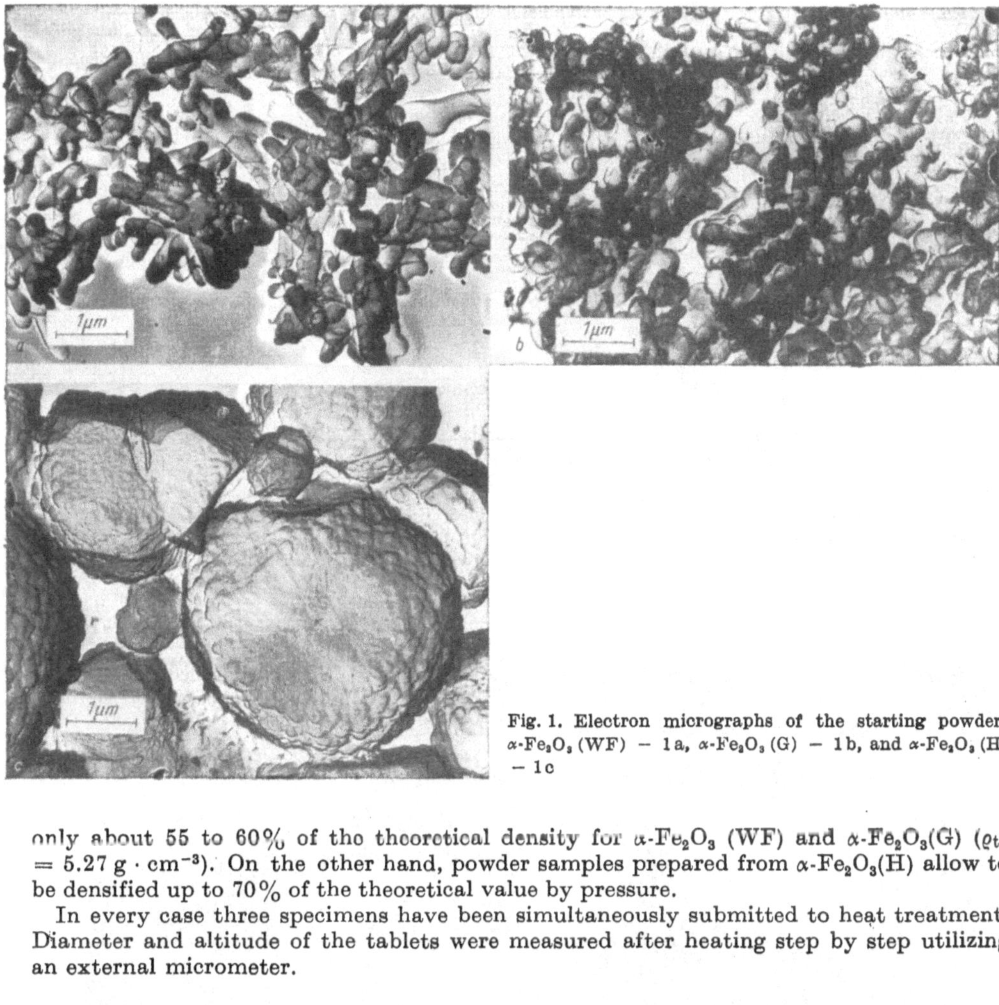

Fig. 1. Electron micrographs of the starting powders α-Fe_2O_3 (WF) – 1a, α-Fe_2O_3 (G) – 1b, and α-Fe_2O_3 (H) – 1c

only about 55 to 60% of the theoretical density for α-Fe_2O_3 (WF) and α-Fe_2O_3(G) ($\varrho_{th} = 5.27\ g \cdot cm^{-3}$). On the other hand, powder samples prepared from α-Fe_2O_3(H) allow to be densified up to 70% of the theoretical value by pressure.

In every case three specimens have been simultaneously submitted to heat treatment. Diameter and altitude of the tablets were measured after heating step by step utilizing an external micrometer.

3. Results

Figure 2 shows the results of sintering at different temperatures for the tablets formed with α-Fe_2O_3(WF) and α-Fe_2O_3(H) powder starting from the density value ϱ_0. The significant difference of the behaviour of the two types of α-Fe_2O_3 is apparently caused by the different morphologic constitution in close relation with the action of impurities.

Though the density value ϱ_0 of the tablet at starting is significantly lower for α-Fe_2O_3(WF) than for α-Fe_2O_3(H), the density in the final state remains evidently lower in the latter case, while the former material yields within surprisingly short times values $\varrho(t)$ which approximate the theoretical density ϱ_{th}. The photographs shown in Figure 3 confirm these results. Sintering of α-Fe_2O_3(H) at 1200 °C yields porous ceramic of small grain size with relatively low compactness. Figure 3c shows a surface of many pores. On the other hand, α-Fe_2O_3(WF) powder compacted under the same pressure provides a relatively dense ceramic (Fig. 3a) α-Fe_2O_3(G) behaves

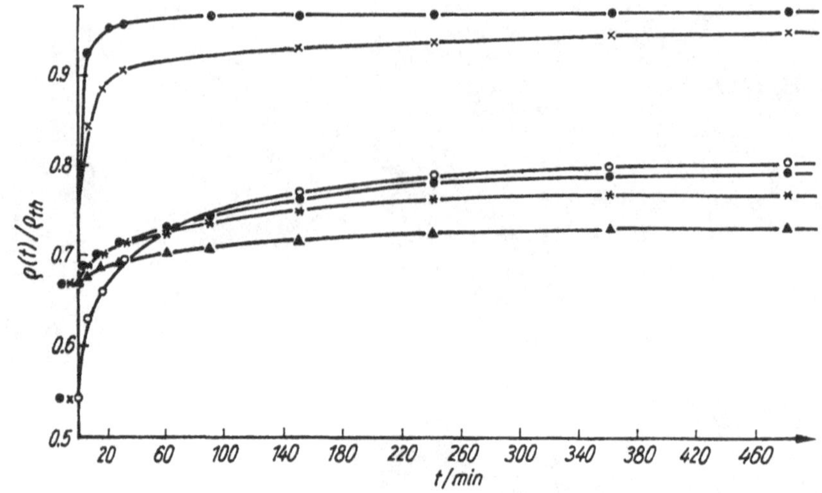

Fig. 2. Density change of α-Fe_2O_3 samples as a function of isothermal heating time (⊗ WF at 1200 °C, × WF at 1050 °C, ○ WF at 900 °C, ● H at 1200 °C, • H at 1050 °C, ▲ H at 900 °C)

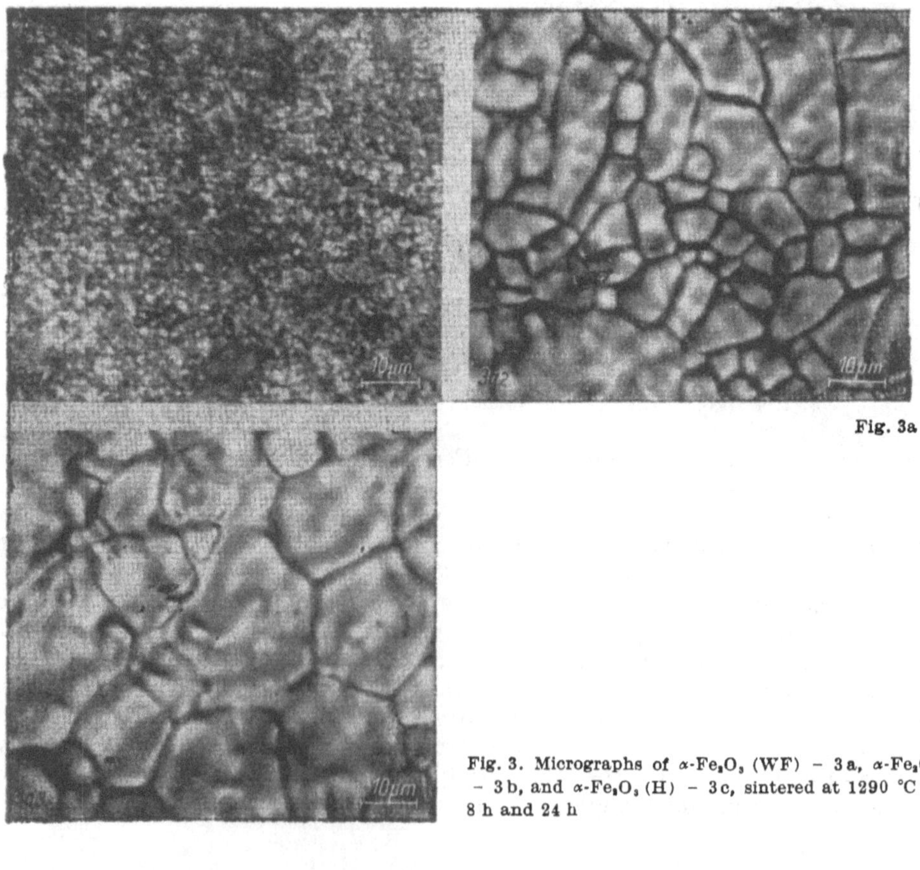

Fig. 3a

Fig. 3. Micrographs of α-Fe_2O_3 (WF) – 3a, α-Fe_2O_3 (G) – 3b, and α-Fe_2O_3 (H) – 3c, sintered at 1290 °C 0.5 h, 8 h and 24 h

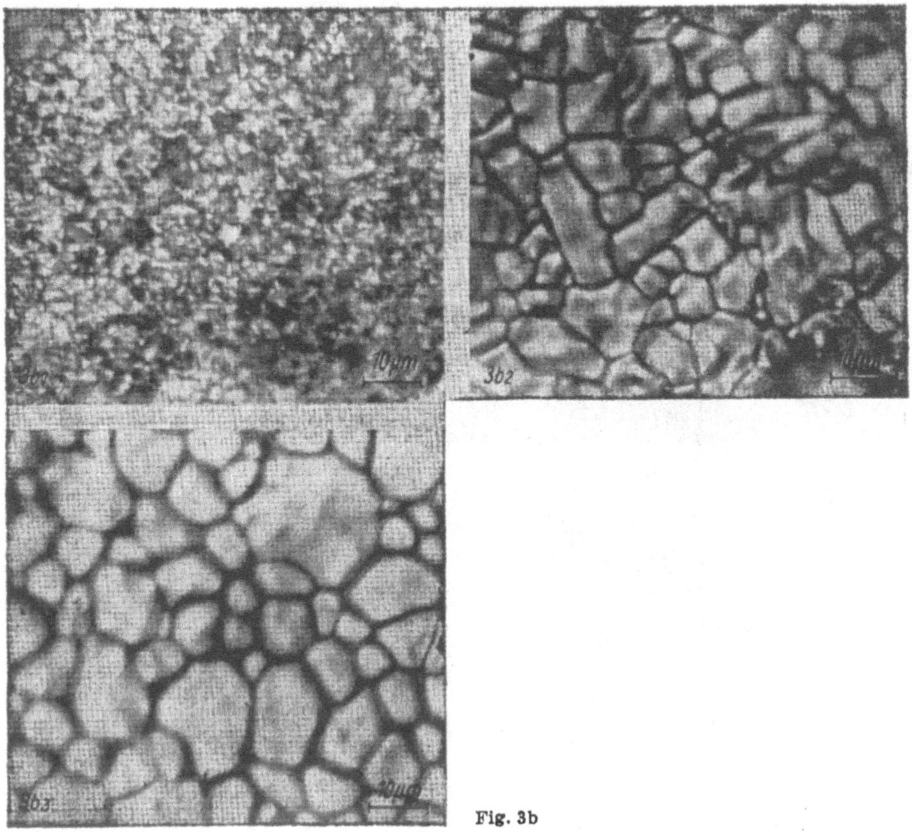

Fig. 3b

Table 2
Sintering data of the α-Fe_2O_3 tablets (temperature in all cases 1200 °C)

α-Fe_2O_3 type	ϱ_0 g/cm³	ϱ_0/ϱ_{th}	ϱ_{480} g/cm³	$\varrho_{480}/\varrho_{th}$
WF	2.870	0.545	5.129	0.973
G	2.900	0.550	4.954	0.940
H	3.531	0.670	4.179	0.793
H (a) (Na)	3.550	0.674	4.148	0.787
H (b) (Al)	3.557	0.675	4.148	0.794
H (c) (Na, Al)	3.520	0.668	4.178	0.793
H (d) (Si)	3.572	0.678	4.161	0.790
H (e) (Si, Mn)	3.559	0.675	4.165	0.790
H (f) (Si, Mn, Al, Na)	3.532	0.670	4.184	0.794

similar to α-Fe_2O_3(WF) also yielding ceramic with only a small content of pores. However, for these two ruws inhomogeneous grain growth is found which becomes more and more evident with sintering time (Fig. 3a and 3b).

The sintering curve of specimens prepared from α-Fe_2O_3(G) for 1200 °C is shown in Figure 4 together with the curves of α-Fe_2O_3(WF) and α-Fe_2O_3(H) for comparison taken from Figure 2. In accordance with expectance from Figure 2 and Table 2 the similarity of the shrinkage dependence on time of samples of Fe_2O_3(WF) and (G) is evident.

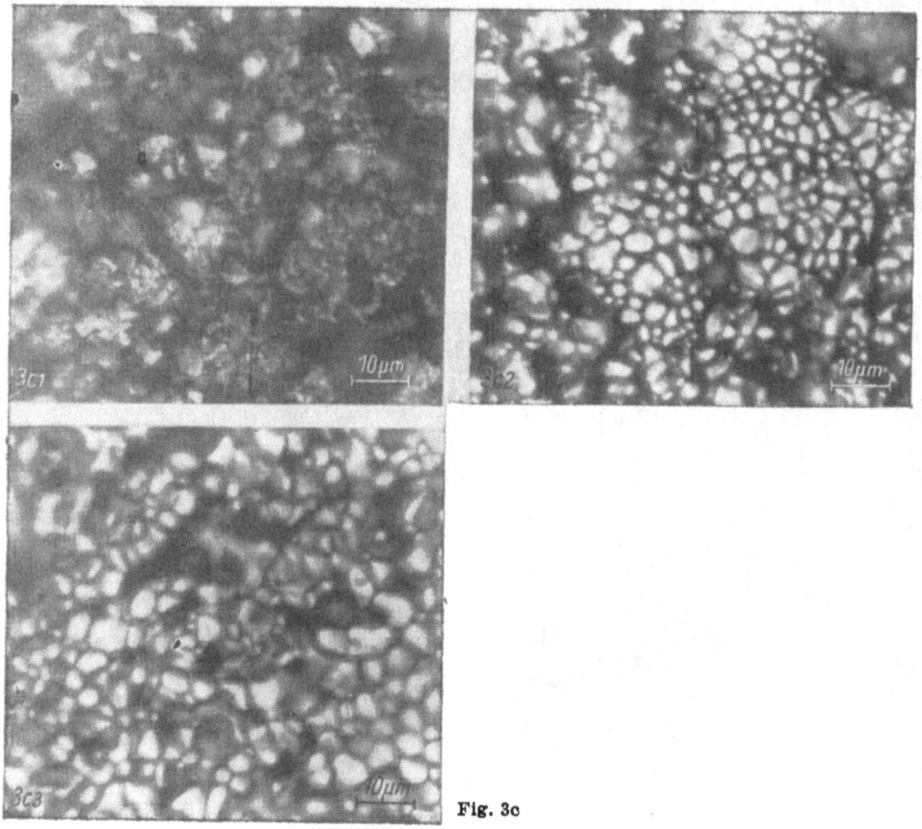

Fig. 3c

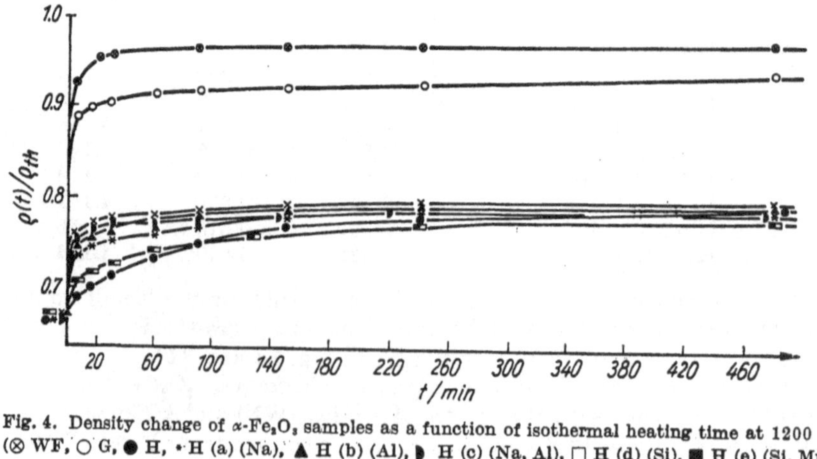

Fig. 4. Density change of α-Fe_2O_3 samples as a function of isothermal heating time at 1200 °C (⊗ WF, ○ G, ● H, • H (a) (Na), ▲ H (b) (Al), ◗ H (c) (Na, Al), □ H (d) (Si), ■ H (e) (Si, Mn), × H (f) (Si, Mn, Al, Na))

Furthermore, Figure 4 shows the curves of shrinkage of the α-Fe_2O_3(H) tablets which were impregnated with different dopants in order to simulate the influence of the impurity content of α-Fe_2O_3(WF) and α-Fe_2O_3(G) on sintering behaviour. As can be seen from the figure there is a significant increase in the shrinkage rate during the first minutes quite similar to the curves of α-Fe_2O_3(WF) and α-Fe_2O_3(G). However the final state remains below that of the last two oxides, which obviously indicates an effect of morphologic powder consitution, real structure memory from origin.

4. Discussion

The results show that the three types of α-Fe_2O_3 have a different sintering behaviour. The powders α-Fe_2O_3(WF) and α-Fe_2O_3(G) which are of relatively small particle size in the beginning, show, in contrast to α-Fe_2O_3(H) with the larger grains, an enormous grain growth. One reason for this is obviosly the essentially more unfavourable granulometric situation of α-Fe_2O_3(H). α-Fe_2O_3(WF)- and α-Fe_2O_3(G)-tablets have a relatively narrow grain size distribution and similar grain shapes yielding relatively small pores after densification by pressure at room temperature. So one finds that the first sintering stage in which the contacts are formed and the necks are rounded is a very rapid one for these products. Grain boundary diffusion is suggested to be favoured in this case.

Although ϱ_0 is considerably greater for α-Fe_2O_3(H) than for the two other oxides the grain boundary diffusion is not so effective for the shrinkage in this case, i.e., for the reduction of the specific surface area energy. Obviously, surface area diffusion predominates yielding a porous ceramic. There is a residual porosity of 20% after sintering and with it an inhibition of grain growth in contrast to α-Fe_2O_3(WF) and α-Fe_2O_3(G). Presumably, the real structure of α-Fe_2O_3(H) prepared by hydrothermal synthesis involves a more ordered state with less internal free enthalpy because of a lower quantity of structural defects. Therefore, the part of volume diffusion becomes reduced. For α-Fe_2O_3(WF) and α-Fe_2O_3(G) a higher value of the internal free enthalpy caused by a larger amount of defects has to be taken into account as a result of the topotactic transformation from α-FeOOH (GÜNTHER, OSWALD). Obviously volume diffusion also plays an important role in the mechanisms of sintering for these two oxides (WATARI et al.).

DE WIT, BROERSMA and STROBAND proposed an oxygen defect phase, i.e. high concentration of defects in the surface near range of the Fe_2O_3-grains at temperatures higher than 950 °C. These defects should be effective for the diffusion processes in all three oxides, but in the case of α-Fe_2O_3(H) they obviously do not play such an important role as in the two others.

Moreover, the influence of impurities on the sintering behaviour has to be taken into consideration. Possibly, the impurity content of α-Fe_2O_3(WF) and α-Fe_2O_3(G) inhibits the mobility of vacancies at the surface which results in a decrease of surface area diffusion and, on the other hand, grain boundary diffusion becomes supported. α-Fe_2O_3(H) is of significantly higher impurity yielding such a porous structure as shown in Figure 3c.

Furthermore, it is known that the formation of liquid phases facilitates the sintering process favouring inhomogeneous grain growth as seen in Figure 3a and 3b. Frequently an extremely small amount of an eutectic melt seems to be sufficient for rearrangement of the grains yielding a denser sample with an increased part of grain boundaries. In consequence of that rapid shrinkage takes place which is impressively shown in the curves of Figure 2 and 4.

The phase diagrams of oxide systems with Na_2O for instance often show eutectics with melting points considerably lower than 1200 °C, e.g., at 1135 °C $(Na_2O)_{0.37} \cdot (Fe_2O_3)_{0.63}$ (KNICK, KOHLMEYER), at 727 °C $(Al_2O_3)_{0.07}(Fe_2O_3)_{0.02}(Na_2O)_{0.29}(SiO_2)_{0.62}$ (BAILEY, SCHAIRER) and at 767 °C $(Al_2O_3)_{0.08}(Na_2O)_{0.20}(SiO_2)_{0.72}$ (SCHAIRER, BOWEN).

The experiments with the dopants Si, Mn, Al, and Na for α-Fe_2O_3(H) shown in Figure 4 indicate a similar behaviour. However, the sintering properties of α-Fe_2O_3 prepared from α-FeOOH cannot be achieved, which indicates the important role of the real structure of the grains and the morphologic constitution of the powders which are introduced into a sintering process.

References

BAILEY, D. K., SCHAIRER, J. F.: J. Petrol. **7**, 114 (1966)
COBLE, R. L.: J. Amer. Ceram. Soc. **41**, 55 (1958)
DE WIT, J. H. W., BROERSMA, A. F., STROBAND, M.: J. Solid State Chem. **37**, 242 (1981)
DE WIT, J. H. W., GEUS, J. W.: Sci. Ceram. **12**, 281 (1983)
FELTZ, A., MARTIN, A.: Reactivity of Solids **2**, 291 (1987) (a)
FELTZ, A., MARTIN, A.: Reactivity of Solids **2**, 307 (1987) (b)
FELTZ, A., MARTIN, A.: Z. anorg. allg. Chem. **540/541**, 26 (1986) (c)
GÜNTHER, J. R., OSWALD, H. R.: Bull. Inst. Chem. Res. Kyoto Univ. **53**, 249 (1975)
IGARASHI, H., KUCZYNSKI, G. C., ENDO, J.: Mater. Sci. Res. **6**, 323 (1973)
KNICK, R., KOHLMEYER, E. J.: Z. anorg. allg. Chem. **244**, 67 (1940)
KÖNIG, U.: Techn. Mitt. Krupp Forsch. Ber. **32**, 75 (1974)
KÖNIG, U.: Ber. Dtsch. Keram. Ges. **55**, 220 (1978)
PFAFF, G., WINDE, H., FELTZ, A., DOLZMANN, R., VOIGT, B., NAUBER, P.: GDR patent 2945891 (1986)
SCHAIRER, J. F., BOWEN, N. L.: Amer. J. Sci. **254**, 129 (1956)
VOIGT, B., GÖBLER, A.: Cryst. Res. Technol. **21**, 1177 (1986)
WATARI, F., DELAVIGNETTE, P., VAN LANDUYT, J., AMELINKX, S.: J. Solid State Chem. **29**, 137 (1979)

(Received July 1, 1987)

Authors' address:

Dr. Gerhard PFAFF, Prof. Dr. A. FELTZ, U. UMPFENBACH
Friedrich-Schiller-Universität Jena, Sektion Chemie
Am Steiger 3
Jena
DDR-6900

Cryst. Res. Technol.	23	1988	1	25–33

H. Görls, B. Schulz, U. Rosenthal, W. Schulz

Zentralinstitut für physikalische Chemie der Akademie der Wissenschaften der DDR, Berlin, and
Zentralinstitut für organische Chemie der Akademie der Wissenschaften der DDR, Rostock

Crystal and Molecular Structure of (2-3-η-2-Butyne-1,4-diol)-bis-(triphenylphosphan)-nickel(O) $Ni\{[(C_6H_5)_3P]_2\ (C_4H_6O_2)\}$

Dedicated to Professor Hermann Neels on the occasion of his 75th birthday

The crystal and molecular structure of (2-3-η-2-Butyne-1,4-diol)-bis-(triphenylphosphan)-nickel(O) has been determined by X-ray structure analysis. It crystallizes in the tetragonal space group $I\bar{4}$ with the cell parameters $a = b = 22.277(3)$, and $c = 15.118\ (4)$ Å. The structure was solved by the heavy atom method and refined to $R = 0.0716$. The co-ordination geometry about the nickel atom is trigonal-planar. Each molecule is hydrogen bonded to two neighbours. The result is an eight-membered oxygen ring.

Die Kristall- und Molekülstruktur von (2-3-η-2-Butin-1,4-diol)-bis-(triphenylphosphan)-nickel(O) wurde mit Hilfe der Röntgenkristallstrukturanalyse bestimmt. Die Verbindung kristallisiert in der tetragonalen Raumgruppe $I\bar{4}$ mit den Zellparametern $a = b = 22.277(3)$ und $c = 15.118(4)$ Å. Die Struktur wurde durch die Schweratommethode gelöst und bis zum R-Wert von 0.0716 verfeinert. Die Koordinationsgeometrie des Nickels ist trigonal-planar. Jedes Molekül ist über Wasserstoffbrücken verbunden. Das Ergebnis ist ein Sauerstoffring.

1. Introduction

Complexes of the type (ligand)$_2$Ni(alkyne) can act as catalysts for the cyclooligomerization of some alkynes and acetylene itself (Jolly et al. 1975, 1982). But there are only very few literature reports on effective cyclooligomerization of functionalized alkynes (Schulz, Rosenthal 1985, 1986).

So some of us investigated as catalysts well defined Ni(O)-complexes, prepared by reactions of (ligand)$_2$Ni(C_2H_4), (ligand)$_4$Ni or (cod)$_2$Ni/2 ligands with functionalized alkynes (Rosenthal 1981, 1984; Rosenthal, Schulz 1985) and found that the different complexation of alkynes is influenced by electronic properties of ligands and alkyne substituent as shown by interpretation of infrared spectra (Rosenthal, Schulz 1987). The degree of complexation is connected with activation of complexed alkynes and illustrated by some reactions, which are different for acceptor- and donor-ligands (Rosenthal, Schulz 1987).

In that investigations it was found in 1981 the first example of a nickel(O)complex with 2-butyne-1,4-diol(ph$_3$P)$_2$Ni(HOCH$_2$–C$_2$–CH$_2$OH) (I) (Rosenthal 1981), which gave suitable crystals for an X-ray structural investigation:

$$(ph_3P)_2Ni(C_2H_4) + HOCH_2\text{–}C_2\text{–}CH_2OH \xrightarrow[-C_2H_4]{} (ph_3P)_2Ni(HOCH_2\text{–}C_2\text{–}CH_2OH) \quad \text{(I)}$$

Complex (I) is characterized by a strong band at 1805 cm^{-1} in the infrared spectra.

To gain informations about the complex structure, to compare the complexation of alkyne and to investigate the special influence of hydroxy-groups a structural investigation seemed necessary.

2. Experimental

For the X-ray diffraction a light yellow crystal of needle shape was used. The intensities of 4949 independent reflections were measured on an Enraf-Nonius CAD-4 diffractometer in a range up to $2\theta = 44°$, using Mo-Kα radiation and an $\omega/2\theta$ scan mode. Corrections for Lorentz and polarisation effects were carried out. No absorption and no extinction corrections were applied. A summary of the crystallographic data are listed in Table 1.

Table 1
Crystal data

crystal morphology	light-yellow needles
chemical formula	$C_{40}H_{36}NiO_2P_2$
molecular weight	669.39 g/mol
crystal system	tetragonal
space group	$I\bar{4}$
cell constants	$a = 22.277(3)$ Å
	$b = 22.277(3)$ Å
	$c = 15.118(4)$ Å
volume	7502.4 Å^3
number of formula units Z	8
density (calculated) d_c	1.185 g/cm^3
absorption coefficient μ(MoK$_\alpha$)	6.317 cm^{-1}
unique data measured	4949
observed data with $I \geqq 1.0\sigma$ (I)	4229
R	0.0716
R_W	0.080
rest electron density	0.842 e/Å^3

3. Structure analysis

The structure was solved by the heavy atom method. The three-dimensional R-value field was calculated from 80 observed reflections with the highest F_0-quantity, using the program POSIT (KRETSCHMAR et al.). The R-value minimum showed position of the nickel atom. The position of other non-hydrogen atoms were determined from successive difference Fourier syntheses of electron density. Fullmatrix least-squares refinement of atomic coordinates and anisotropic temperature factors using unit weights converged at $R = 0.09$. After generation of the H atom positions without H atoms from the two hydroxygroups from assumed geometries the refinement of the non-hydrogen coordinates with anisotropic vibrational parameters led to $R = 0.08$. In the final stage of the least-squares procedure H positions were refined with fixed isotropic thermal parameters. This converged with a final $R = 0.0716$. The resulting atomic parameters are listed in Table 2. All calculation out of POSIT were performed on a PDP 11/34 minicomputer with the ENRAF-NONIUS SDP program package.

4. Discussion

Perspective view of (2-3-η-2-Butyne-1,4-diol)-bis-(triphenylphosphan)-nickel(O) with the numbering scheme is given in Figure 1. Figure 2 shows a stereo plot of the molecule. Bond distances and bond angles are given in Table 3.

Table 2
Positional parameters and equivalent isotropic temperature factors (a.s.d.'s in parantheses)

Atom	x/a	y/b	z/c	B_{eq}
Ni1	0.20694(5)	0.25503(5)	0.18698(8)	3.85(2)
P1	0.3037(1)	0.2607(1)	0.1972(2)	3.95(5)
P2	0.1806(1)	0.1623(1)	0.1706(2)	4.12(5)
O1	0.0425(3)	0.3460(3)	0.2398(7)	8.7(3)
O2	0.1363(4)	0.4264(3)	0.2578(8)	10.3(3)
C1	0.0672(4)	0.2900(5)	0.2093(9)	6.3(3)
C2	0.1352(4)	0.2939(4)	0.2056(7)	4.5(2)
C3	0.1768(4)	0.3300(4)	0.2085(7)	4.7(2)
C4	0.1889(5)	0.3971(4)	0.2275(9)	6.8(3)
C5	0.3320(4)	0.3372(4)	0.2156(6)	4.2(2)
C6	0.3311(5)	0.3769(4)	0.1456(7)	5.4(2)
C7	0.3465(5)	0.4351(5)	0.1551(8)	6.0(3)
C8	0.3622(6)	0.4554(5)	0.2390(9)	7.8(3)
C9	0.3640(7)	0.4157(5)	0.308(1)	10.2(4)
C10	0.3482(6)	0.3569(5)	0.2992(9)	7.3(3)
C11	0.3568(4)	0.2376(4)	0.1117(7)	4.5(2)
C12	0.4182(4)	0.2482(5)	0.1216(7)	5.4(3)
C13	0.4595(6)	0.2342(6)	0.056(1)	8.3(4)
C14	0.4372(6)	0.2042(6)	−0.025(1)	8.5(4)
C15	0.3730(6)	0.1937(5)	−0.0342(8)	6.9(3)
C16	0.3325(5)	0.2114(4)	0.0338(7)	5.9(3)
C17	0.3254(4)	0.2173(4)	0.2961(6)	4.1(2)
C18	0.2832(5)	0.2213(5)	0.3665(7)	5.3(2)
C19	0.2929(5)	0.1869(5)	0.4422(9)	6.9(3)
C20	0.3401(6)	0.1474(6)	0.4458(9)	8.3(3)
C21	0.3807(6)	0.1405(6)	0.372(1)	8.3(4)
C22	0.3712(5)	0.1792(5)	0.3006(8)	6.4(3)
C23	0.1168(4)	0.1550(4)	0.0972(6)	3.8(2)
C24	0.1109(5)	0.1992(5)	0.0324(7)	5.4(3)
C25	0.0645(5)	0.1962(5)	−0.0309(7)	6.0(3)
C26	0.0255(5)	0.1496(6)	−0.0285(7)	6.6(3)
C27	0.0300(4)	0.1042(5)	0.0368(8)	6.0(3)
C28	0.0777(4)	0.1088(5)	0.1003(7)	5.0(2)
C29	0.1533(5)	0.1240(4)	0.2718(6)	4.6(2)
C30	0.1320(5)	0.1611(5)	0.3390(7)	5.3(2)
C31	0.1064(6)	0.1357(6)	0.4137(8)	6.8(3)
C32	0.1009(6)	0.0739(6)	0.4203(8)	7.8(4)
C33	0.1255(6)	0.0359(6)	0.3535(9)	8.5(4)
C34	0.1513(5)	0.0626(5)	0.2770(7)	5.8(3)
C35	0.2327(5)	0.1100(4)	0.1242(8)	5.3(3)
C36	0.2345(5)	0.0970(5)	0.0314(8)	6.6(3)
C37	0.2814(6)	0.0610(5)	−0.003(1)	9.0(4)
C38	0.3284(7)	0.0351(6)	0.055(2)	12.5(5)
C39	0.3265(6)	0.0485(6)	0.145(1)	10.8(5)
C40	0.2792(5)	0.0859(5)	0.179(1)	7.6(3)
H01	0.0239	0.3880	0.2626	
H02	0.0961	0.4487	0.2808	
H11	0.0500	0.2803	0.1440	
H12	0.0542	0.2546	0.2542	
H41	0.2232	0.4007	0.2775	
H42	0.2040	0.4185	0.1674	

Table 2 (cont.)

Atom	x/a	y/b	z/c	B_{eq}
H61	0.3177	0.3609	0.0811	
H71	0.3464	0.4651	0.0991	
H81	0.3728	0.5020	0.2496	
H91	0.3786	0.4316	0.3715	
H101	0.3481	0.3269	0.3553	
H121	0.4341	0.2679	0.1824	
H131	0.5064	0.2450	0.0640	
H141	0.4679	0.1902	−0.0761	
H151	0.3559	0.1724	−0.0932	
H161	0.2847	0.2050	0.0264	
H181	0.2445	0.2504	0.3613	
H191	0.2630	0.1912	0.4982	
H201	0.3469	0.1209	0.5048	
H211	0.4162	0.1075	0.3726	
H221	0.4027	0.1780	0.2463	
H241	0.1422	0.2361	0.0306	
H251	0.0601	0.2306	−0.0807	
H261	−0.0097	0.1469	−0.0773	
H271	−0.0014	0.0673	0.1509	
H281	0.0823	0.0748	0.1509	
H301	0.1353	0.2093	0.3329	
H311	0.0906	0.1641	0.4667	
H321	0.0779	0.0543	0.4762	
H331	0.1244	−0.0123	0.3609	
H341	0.1689	0.0351	0.2241	
H361	0.2002	0.1145	−0.0120	
H371	0.2831	0.0520	−0.0731	
H381	0.3631	0.0065	0.0287	
H391	0.3604	0.0308	0.1888	
H401	0.2788	0.0962	0.2486	

The triphenylphosphan ligands are arranged like a propeller. This is the arrangement with a minimum for steric hindrance. The Ni1–P1 and Ni1–P2 bond lengths are 2.166(1) Å, and 2.161(1) Å, respectively. The average values of the phosphorus-carbon and the carbon-carbon distances in the phenyl rings amount to 1.827(4) Å, and 1.402(8) Å, respectively. This bond lengths are typical for the triphenylphosphan nickel framework (WALTHER et al.; FERGUSON et al.; KEIM et al.; COOK et al.). But the carbon-carbon distances of the phenyl rings ranged from 1.328(6) Å to 1.485(10) Å. The high differences from the theoretical value of 1.39 Å (BROWN) originated on the strength of the high thermal vibrations of the atoms C8, C14, C20, C26, C32 and their vicinal carbon atoms (see B_{eq} in Table 2).

The alkyne ligand is coordinated "side-on" about the carbon-carbon triple bond. The coordinated alkyne is no longer linear but displays a cis geometry. The deviation from linearity (C1–C2–C3 = 142.1(5)°, and C2–C3–C4 = 140.5(4)°, respectively) is typical for η-coordinated alkynes. The dihedral angle is 9.7° between the planes Ni1, P1, P2 and Ni1, C2, C3. The coordination about the nickel atom is trigonal-planar. This is characteristic of nickel-alkyne complexes (PÖRSCHKE et al.; DICKSON, IBERS; APOTECHER) with the exception of a complex with a silyl-substitute alkyne (ROSENTHAL et al. 1987).

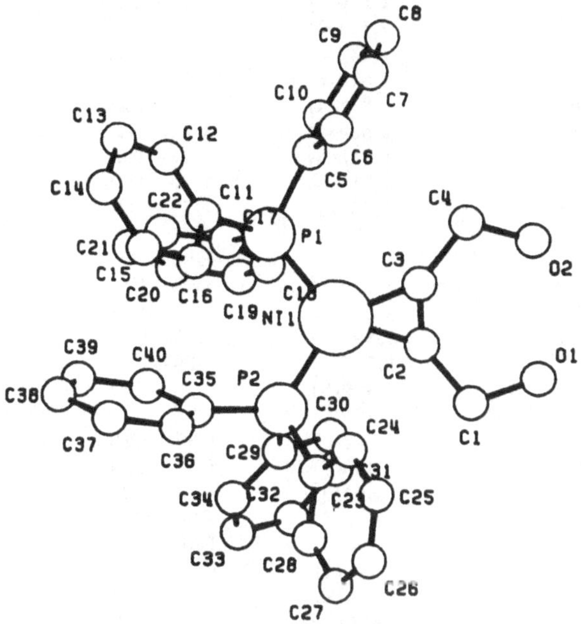

Fig. 1. The (2-3-η-2-Butyne-1,4-diol) bis-(triphenylphosphan) nickel(O) molecule with numbering scheme

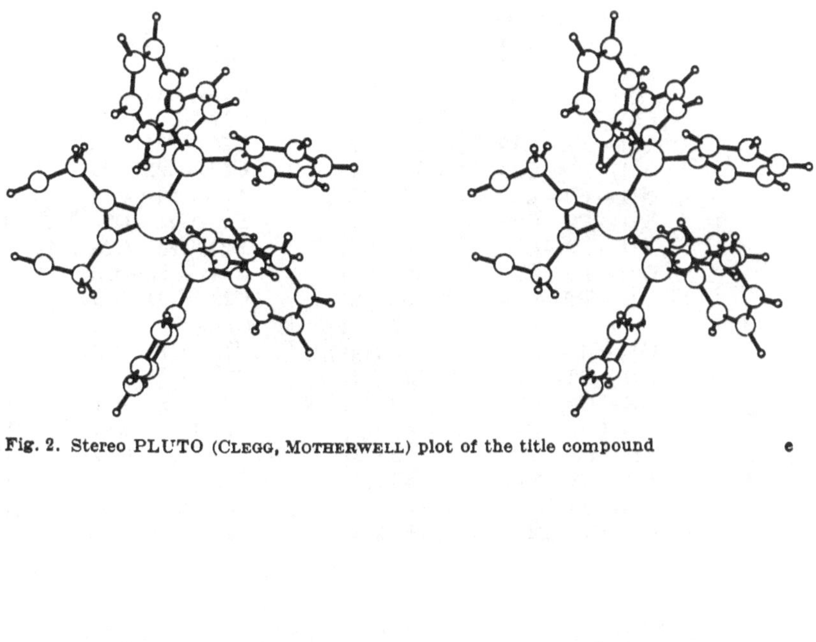

Fig. 2. Stereo PLUTO (CLEGG, MOTHERWELL) plot of the title compound

Table 3
Interatomic distances (A) and bond angles (0)

Ni1–P1	2.166(1)	C15–C16	1.424(7)
Ni1–P2	2.161(1)	C17–C18	1.423(6)
Ni1–C2	1.839(5)	C17–C22	1.328(6)
Ni1–C3	1.829(5)	C18–C19	1.395(7)
P1–C5	1.838(4)	C19–C20	1.373(8)
P1–C11	1.825(4)	C20–C21	1.439(9)
P1–C17	1.845(4)	C21–C22	1.403(8)
P2–C23	1.810(4)	C23–C24	1.396(6)
P2–C29	1.853(5)	C23–C28	1.349(6)
P2–C35	1.789(5)	C24–C25	1.410(7)
O1–C1	1.438(6)	C25–C26	1.354(8)
O2–C4	1.416(6)	C26–C27	1.416(8)
C1–C2	1.519(6)	C27–C28	1.435(6)
C2–C3	1.228(6)	C29–C30	1.394(6)
C3–C4	1.546(6)	C29–C34	1.371(6)
C5–C6	1.379(6)	C30–C31	1.386(7)
C5–C10	1.387(7)	C31–C32	1.387(8)
C6–C7	1.348(6)	C32–C33	1.426(9)
C7–C8	1.390(8)	C33–C34	1.423(8)
C8–C9	1.364(9)	C35–C36	1.434(7)
C9–C10	1.361(7)	C35–C40	1.429(7)
C11–C12	1.398(6)	C36–C37	1.418(10)
C11–C16	1.420(7)	C37–C38	1.48(2)
C12–C13	1.385(8)	C38–C39	1.385(13)
C13–C14	1.485(10)	C39–C40	1.439(9)
C14–C15	1.454(9)		
P1–Ni1–P2	109.53(4)	C11–C12–C13	122.4(6)
P1–Ni1–C2	145.8(1)	C12–C13–C14	117.9(6)
P1–Ni1–C3	107.4(1)	C13–C14–C15	118.8(6)
P2–Ni1–C2	103.4(1)	C14–C15–C16	120.6(6)
P2–Ni1–C3	142.5(1)	C11–C16–C15	118.1(5)
C2–Ni1–C3	39.1(2)	P1–C17–C18	113.6(3)
Ni1–P1–C5	113.9(1)	P1–C17–C22	125.2(4)
Ni1–P1–C11	125.3(2)	C18–C17–C22	120.6(5)
Ni1–P1–C17	106.7(1)	C17–C18–C19	118.5(5)
C5–P1–C11	98.5(2)	C18–C19–C20	120.2(6)
C5–P1–C17	105.9(2)	C19–C20–C21	121.4(6)
C11–P1–C17	104.9(2)	C20–C21–C22	115.8(6)
Ni1–P2–C23	111.7(1)	C17–C22–C21	123.2(6)
Ni1–P2–C29	115.8(2)	P2–C23–C24	116.1(3)
Ni1–P2–C29	115.8(2)	P2–C23–C24	116.1(3)
Ni1–P2–C35	119.4(2)	P2–C23–C28	123.7(4)
C23–P2–C29	102.0(2)	C24–C23–C28	120.1(5)
C23–P2–C35	102.1(2)	C23–C24–C25	120.8(5)
C29–P2–C35	103.6(2)	C24–C25–C26	119.2(5)
O1–C1–C2	110.1(4)	C25–C26–C27	121.3(5)
Ni1–C2–C1	147.7(4)	C26–C27–C28	118.0(5)
Ni1–C2–C3	70.0(3)	C23–C28–C27	120.5(5)
C1–C2–C3	142.1(5)	P2–C29–C30	116.1(3)
Ni1–C3–C2	70.9(3)	P2–C29–C34	121.2(4)
Ni1–C3–C4	148.5(3)	C30–C29–C34	122.6(5)
C2–C3–C4	140.5(4)	C29–C30–C31	119.5(5)
O1–C4–C3	111.2(4)	C30–C31–C32	120.0(6)

Table 3 (cont.)

P1−C5−C6	118.3(3)	C31−C32−C33	120.3(6)
P1−C5−C10	121.4(4)	C32−C33−C34	118.9(6)
C6−C5−C10	120.0(4)	C29−C34−C35	118.5(5)
C5−C6−C7	122.1(5)	P2−C35−C36	122.2(4)
C6−C7−C8	118.3(5)	P2−C35−C40	119.3(4)
C7−C8−C9	119.4(5)	C36−C35−C40	118.0(5)
C8−C9−C10	123.0(6)	C35−C36−C37	119.6(8)
C5−C10−C9	117.2(6)	C36−C37−C38	121.4(9)
P1−C11−C12	120.7(4)	C37−C38−C39	118.5(7)
P1−C11−C16	117.2(3)	C38−C39−C40	119.6(8)
C12−C11−C16	122.1(5)	C35−C40−C39	122.9(7)

Another nickel alkynediol complex which has been structurally characterized is the compound (2-3-η-2-butyne-1,4-diol) N,N′-ethanediylidene bis(2,6-diisopropylaniline) nickel(O)-II- (Diercks et al.). In both nickel alkynediol frameworks the nickel-carbon bond lengths are in good agreement within 3σ

(I: Ni1–C2 = 1.839(5) Å, Ni1–C3 = 1.829(5) Å;

II: Ni1–C2 = Ni1–C3 = 1.840(3) Å).

The infrared spectra (solid matter-Nujol technique) show a strong band for the coordinated triple bond at 1805 cm^{-1} (I), and 1811 cm^{-1} (II), respectively. That means the IR-bands are nearly the same for the two compounds. Therefore it was expected in both complexes the carbon-carbon triple bond lengths and angles are equal too. But the results of the crystal structures show the bond distances C2–C3 (I: 1.228(6) Å; II: 1.273(7) Å), and the angles C1–C2–C3, and C2–C3–C4 (I: 142.2(5)° 140.5(4)°; II: 152.4(3)°, 152.4(3)°) are significantly different from each other. The carbon-carbon distance of 1.273(7) Å for II is intermediate between that for normal C≡C (1.20 Å) and C=C bonds (1.34 Å). Contrary to compound II the distance of 1.228(6) Å reflect in the title compound a high triple bond part.

The molecular packing arrangement along c is illustrated in Figure 3. The diagram indicated the structure stabilized by intra- and intermolecular OH ... O hydrogen

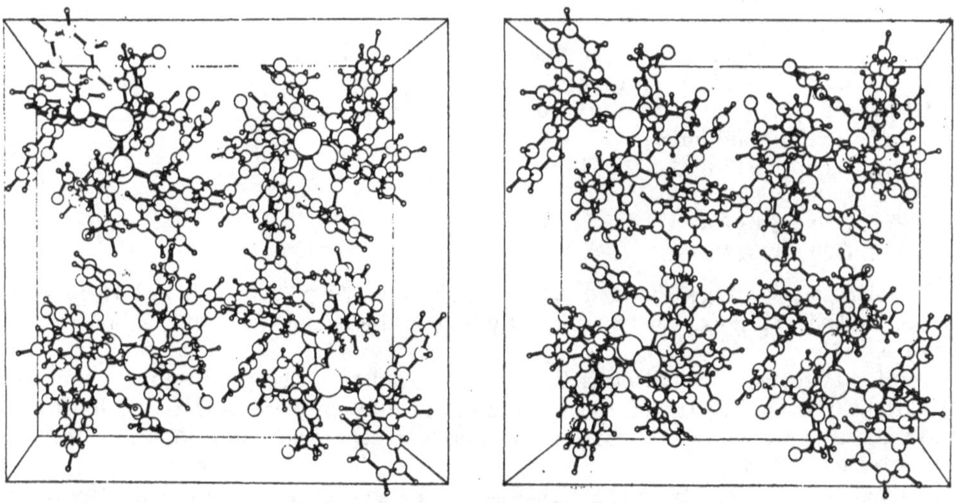

Fig. 3. Molecular packing in the crystal viewed along the c-axis

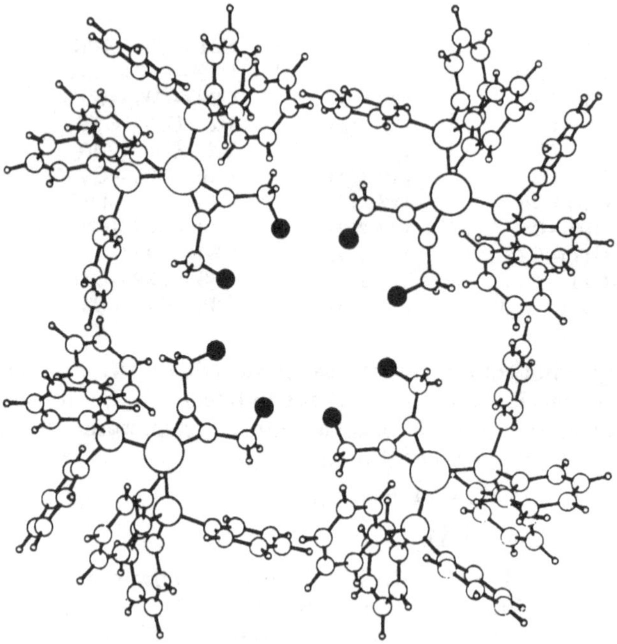

Fig. 4. Stereo plot of the four molecules which formed the eight-membered oxygen ring (oxygen atoms are indicated by full circles)

bonds. Always the intermolecular hydrogen bonds are formed between four molecules (Fig. 4). The result is an eight-membered oxygen ring.

The thermal vibration of the oxygen atoms are very high (see B_{eq} in Table 2) therefore no attempt was made to locate the hydrogen atoms from the difference Fourier map. The presence of OH ... O bonds can be inferred from a consideration of intra- and interatomic distances within of the eight-membered oxygen ring:

a) intramolecular bond lengths

O1 ... O2 2.766(6) Å

O1^{i} ... O2^{ii} 2.766(6) Å (i ; ii : $.5 - y$, $.5 + x$, $.5 - z$)

O1^{iii} ... O2^{iv} 2.766(6) Å (iii; iv: $-x$, $1 - y$, z)

O1^{v} ... O2^{vi} 2.766(6) Å (v; vi: $.5 - y$, $.5 - x$, $.5 - z$)

b) intermolecular bond lengths

O1 ... O2^{vi} 2.617(6) Å

O2 ... O1^{i} 2.617(6) Å

O1^{iii} ... O2^{ii} 2.617(6) Å

O1^{v} ... O2^{iv} 2.617(6) Å.

Each molecule is hydrogen bonded to two neighbours. At present it is known in what respect a relationship exist between the hydrogen bonds and the short carbon-carbon distance.

Complex II formed no hydrogen bonds. Compared to it the platinium(O) complex $Pt[(C_2H_5)_2C(OH)C{\equiv}CC(OH)\ (C_2H_5)_2]_2$ is stabilized by a network of OH ... O hydrogen bonds. But contrary to I the carbon triple bond is stretched to 1.35(3) Å, and 1.36(2) Å, respectively (Dubey).

The leat-square plane from the eight-membered oxygen ring shows that the intermolecular connected atoms arranged alternate downward and upward. The distances from the least-square plane range from ± 0.12 to ± 0.15 Å. The eight-membered ring has a boat conformation.

References

APOTECHER, B.: Dissertation, Bochum 1983
BROWN, M. G.: Trans. Faraday Soc. **55**, 694 (1959)
CLEGG, W., MOTHERWELL, W. D. S.: PLUTO 78. A program for plotting molecular and crystal structures
COOK, C. D., CKUNG HOE KOO, NYBURG, S. C., SHIOMI, M. T.: Chem. Comm. 426 (1967)
DICKSON, R. S., IBERS, J. A.: J. Organomet. Chem. **36**, 191 (1972)
DIERCKS, R., KOPF, J., TOM DIECK, H.: Acta Cryst. **C40**, 363 (1984)
DUBEY, R. J.: Acta Cryst. **B31**, 1860 (1975)
FERGUSON, G., SIEW, P. Y., YALOR, F. J., DESMOND, T. J.: J. Chem. Soc. Dalton Trans. 1981 (1982)
JOLLY, P. W., WILKE, G.: The Organic Chemistry of Nickel, Vol. 2, London 1975
JOLLY, P. W.: Comprehensive Organometallic Chemistry, Vol. 8, New York 1981
KEIM, W., KOWALDT, F. H., GODDARD, R., KRÜGER, C.: Angew. Chem. **90**, 493 (1978)
KRETSCHMAR, R. G., RECK, G., KUTSCHABSKY, L.: Program POSIT, Berlin
PÖRSCHKE, K. R., TSAY, Y. H., KRÜGER, C.: Angew. Chem. **97**, 334 (1985)
PÖRSCHKE, K. R., MYNOTT, R., ANGERMUND, K., KRÜGER, C.: Z. Naturforsch. **40b**, 199 (1985)
ROSENTHAL, U.: Z. anorg. allg. Chem. **482**, 179 (1981)
ROSENTHAL, U.: Z. anorg. allg. Chem. **513**, 151 (1984)
ROSENTHAL, U., SCHULZ, W.: Z. Chem. **25**, 227 (1985)
ROSENTHAL, U., SCHULZ, W.: J. Organomet. Chem. printed
ROSENTHAL, U., SCHULZ, W., GÖRLS, H.: Z. anorg. allg. Chem. printed
SCHULZ, W., ROSENTHAL, U.: Z. Chem. **26**, 177 (1986)
SCHULZ, W., ROSENTHAL, U.: DDR-Pat. C07/C279561-6 (1985)
WALTHER, D., DINJUS, E., GÖRLS, H., SIELER, J., LINDQUIST, O., ANDERSON, L.: J. Organomet. Chem. **286**, 103 (1985)

(Received June 22, 1987)

Authors' addresses:

Dr. H. GÖRLS, B. SCHULZ
Zentralinstitut für physikalische Chemie der
Akademie der Wissenschaften der DDR
Rudower Chaussee 5
Berlin
DDR-1199

Dr. U. ROSENTHAL, Dr. W. SCHULZ
Zentralinstitut für organische Chemie der Akademie der Wissenschaften der DDR
Bereich Komplexkatalyse
Buchbinderstr. 5–6
Rostock
DDR-2500

Cryst. Res. Technol.	23	1988	1	35–39

U. Braun, J. Sieler, R. Richter, I. Leban, L. Golič

Karl-Marx-Universität, Sektion Chemie, and
Edvard-Kardelj-Universität Ljubljana, Abteilung für Chemie und Chemische Technology

Crystal and Molecular Structure of N-(Diethylaminothiocarbonyl)-N'-phenyl-benzamidine

Dedicated to Professor Hermann Neels on the occasion of his 75th birthday

The crystal structure of the title compound has been determined by X-ray diffractometer data collected on a CAD-4 diffractometer. $C_{18}H_{21}N_3S$ crystallizes monoclinic, space group P 2_1/n with $a = 8.939(1)$ Å, $b = 18.992(3)$ Å, $c = 10.416(2)$ Å, $\beta = 97.29(2)$ Å and $Z = 4$. Least-squares refinement gave a value of $R = 0.107$ for 2647 observed reflections. The molecule has a configuration which can be described best by Z, E' and exists in the tautomeric enamine form. The molecules form infinite chains connected by intermolecular N–H ... S interactions, with H ... S distances of 2.46 Å.

Die Kristallstruktur der Titelverbindung wurde durch Röntgenstrukturanalyse bestimmt. Die Daten wurden mit einem CAD-4 Diffraktometer gesammelt. $C_{18}H_{21}N_3S$ kristallisiert monoklin in der Raumgruppe P 2_1/n mit $a = 8{,}939(1)$ Å, $b = 18{,}992(3)$ Å, $c = 10{,}416(2)$ Å, $\beta = 97{,}29(2)°$ und $Z = 4$. Die Verfeinerung nach der Methode der kleinsten Fehlerquadrate ergab einen R-Wert von 0,107 für 2647 beobachtete Reflexe. Das Molekül besitzt eine Konfiguration, die am besten mit Z, E' beschreiben werden kann und liegt in der tautomeren Enamin-Form vor. Die Moleküle bilden unendliche Ketten, die durch intermolekulare N–H ... S-Wechselwirkungen verbunden sind. Die H ... S Abstände betragen 2,46 Å.

Introduction

This structure determination is the third in a series where we are examining N-chalcogenoacylthiourea ligands (Braun et al. a, b). In the present structure the chalcogen atom is substituted by an amino group NR ($R = C_6H_5$) in which the N atom is formally isoelectronic to the chalcogen atom O. As can be seen from the formula two tautomeric forms are possible (I or II).

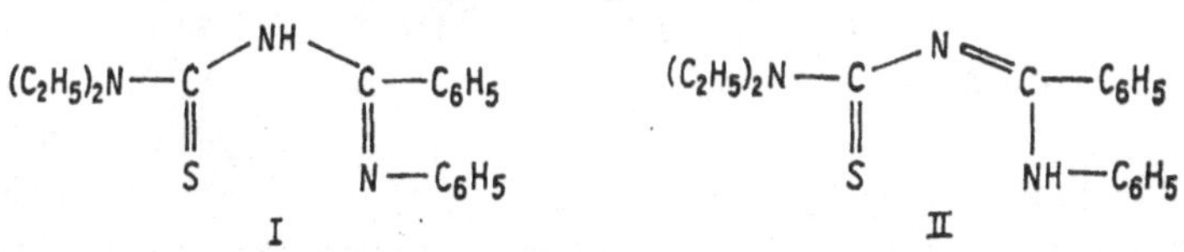

Since we already investigated the copper (Richter et al.) and the nickel chelate (Sieler et al.) of this ligand we were interested in structural features concerning the uncoordinated ligand especially such as NH tautomerism and configuration.

Crystal data

The compound was prepared according to the method of Beyer et al.

Single crystal diffractometry, graphite monochromated MoKα radiation, $\lambda = 0.71069$ Å, space group P 2_1/n $a = 8.939(1)$ Å, $b = 18.992(3)$ Å, $c = 10.416(2)$ Å, $\beta = 97.29(2)°$, $V = 1754.0$ Å^3, number of formula units $C_{18}H_{21}N_3S$ in the unit cell $Z = 4$, calculated density $D_x = 1.18$ g cm^{-3}, linear absorption coefficient $\mu = 1.47$ cm^{-1}, crystal size $0.2 \times 0.1 \times 0.1$ mm^3, yellow prisms, m.p. 101–102 °C (ethanol).

Structure determination

The intensity data were collected on a CAD-4 diffractometer. Threedimensional data (3247 reflections with $2\theta \leq 55°$) were measured using the $\omega/2\theta$ scan technique. 3247 reflections were used for the structure determination, for the least-squares refinement 2647 reflections with $F_0 > 4\sigma(F_0)$ were used. The data were corrected for Lorentz and polarisation effects, an absorption correction was not applied. The structure was solved by using the program SHELX76 (SHELDRICK) using direct methods. All H atoms could be obtained from a difference map. The refinement was performed by use of the program system NRC (AHMED et al.). In the final refinement anisotropic temperature factors for all non-hydrogen atoms and isotropic temperature factors of the attached atom for the hydrogen atoms in their calculated positions, except for

Table 1
Atomic coordinates ($\times 10^4$) and isotropic thermal parameters (Å^2)

	x	y	z	B^*_{eq}
S	2425(2)	2182(1)	3711(1)	2.30(3)
N(1)	1807(5)	1571(3)	5900(5)	2.5(1)
N(2)	4228(6)	1459(4)	5460(5)	3.7(2)
N(3)	−147(5)	1846(3)	7056(5)	2.4(1)
C(1)	1040(6)	2042(3)	6421(5)	2.2(1)
C(2)	2880(6)	1740(3)	5104(6)	2.3(1)
C(3)	4450(11)	777(8)	6577(10)	6.2(4)
C(4)	4721(14)	1230(8)	7531(19)	8.8(6)
C(5)	5484(8)	1556(5)	4706(7)	3.7(2)
C(6)	6351(9)	2198(6)	5097(8)	4.8(3)
C(7)	1402(7)	2808(4)	6475(6)	2.8(2)
C(8)	307(9)	3299(4)	6129(8)	3.6(2)
C(9)	696(12)	4020(4)	6218(9)	4.6(3)
C(10)	2145(11)	4216(4)	6652(10)	5.1(3)
C(11)	3221(11)	3718(6)	7021(10)	5.5(3)
C(12)	2868(8)	3012(5)	6940(8)	3.9(2)
C(13)	−653(6)	1149(3)	7183(6)	2.3(1)
C(14)	−1010(8)	929(4)	8363(7)	3.2(2)
C(15)	−1560(10)	248(4)	8514(8)	3.8(2)
C(16)	−1753(9)	−206(4)	7478(9)	3.7(2)
C(17)	−1422(8)	15(4)	6294(8)	3.4(2)
C(18)	−868(7)	690(4)	6124(6)	2.8(2)

$B^*_{eq} = 3/4\,(a^2B_{11} + b^2B_{22} + c^2B_{33} + 2acB_{13}\cos\beta)$

Table 2
Coordinates of the hydrogen atoms ($\times 10^4$)

	x	y	z		x	y	z
H(31)	3389	512	6578	H(10)	2447	4747	6696
H(32)	5425	484	6448	H(11)	4323	3889	7380
H(51)	6232	1118	4838	H(12)	3709	2628	7208
H(52)	5097	1583	3705	H(14)	1282	1282	9164
H(61)	6071	2673	4933	H(15)	−1804	71	9432
H(62)	6544	2315	6099	H(16)	−2166	−724	7583
H(63)	7369	2297	4700	H(17)	−1575	−336	5506
H(8)	−828	3140	5800	H(18)	−622	852	5194
H(9)	−116	4425	5972	H(1)	−731	2245	7488

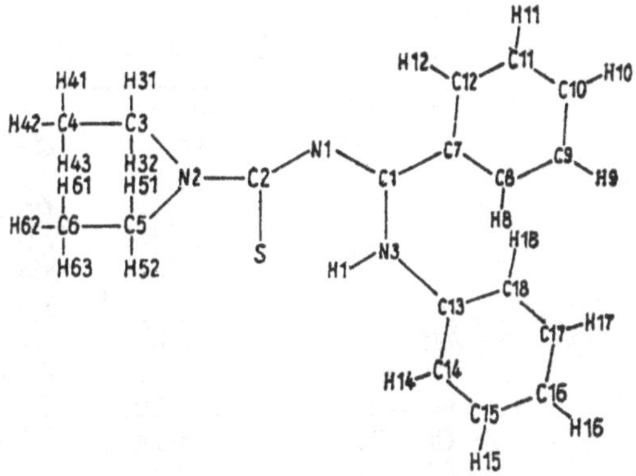

Fig. 1. The atom numbering scheme

H1, H41-43 and H61-63, for which the experimental coordinates were used, were included. The final $R\ (= \sum |F_0| - |F_c|/\sum |F_0|)$ value was 0.107. The positional parameters of the non-hydrogen atoms and hydrogen atoms are given in Table 1 and Table 2, respectively.

Lists of the structure factors are available from the authors on request.

The atom numbering scheme is depicted in Figure 1.

Discussion

Figure 2 shows the molecular structure of the title compound. Bond distances and bond angles are listed in Table 3. The S,C2,N1,C1,N3 unit deviates extremly from planarity whereas the thiourea moiety S,C2,N1,N2 and the C2,N2,C3,C5 units are planar within experimental error indicating sp^2-character of N2.

The molecule adopts a configuration which can be described best as the Z,E′ form. The torsion angles ω_1 and ω_2 (Braun et al. a) have values of 61.0 and −171.1°, respectively.

The hydrogen atom H1 is bonded to the N3 atom. Thus the molecule exists in the tautomeric enamine form which also has been found for the N3 unsubstituted N-(diethylaminothiocarbonyl)-benzamidine (Braun et al. b).

The C–S bond length has a value of 1.681 Å lieing between single (1.81 Å) and double bond (1.61 Å) (Pauling) and corresponds to distances in thioureas (Truter a; Elcombe, Taylor; Dias, Truter) and thioamides (Truter b).

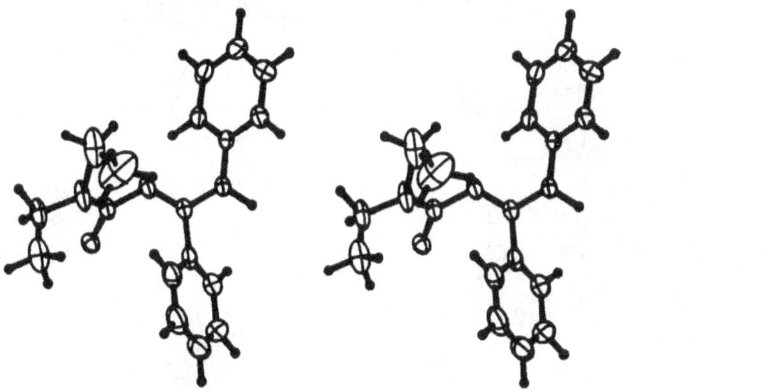

Fig. 2. Stereoscopic view of the molecular structure of the title compound (Johnson)

Table 3
Bond lengths (Å) and bond angles (°)

S–C2	1.681(6)	C1–N1–C2	122.5(5)
N1–C1	1.289(8)	C2–N2–C3	120.8(6)
N1–C2	1.382(7)	C2–N2–C5	122.1(6)
N2–C2	1.327(8)	C3–N2–C5	115.0(6)
N2–C3	1.735(15)	C1–N3–C13	125.3(5)
N2–C5	1.461(9)	N1–C1–C7	124.8(5)
N3–C1	1.370(7)	N1–C1–N3	119.9(5)
N3–C13	1.410(8)	C7–C1–N3	115.2(5)
C1–C7	1.488(9)	S–C2–N1	121.5(4)
C3–C4	1.313(22)	S–C2–N2	123.7(5)
C5–C6	1.474(14)	N1–C2–N2	114.4(5)
C7–C8	1.367(10)	N2–C3–C4	90.7(10)
C8–C9	1.414(12)	N2–C5–C6	111.7(7)
C9–C10	1.369(10)	C1–C7–C8	120.7(6)
C10–C11	1.370(10)	C1–C7–C12	118.4(6)
C11–C12	1.376(14)	C8–C7–C12	120.8(7)
C12–C7	1.395(10)	C7–C8–C9	118.8(7)
C13–C14	1.372(9)	C8–C9–C10	120.0(8)
C14–C15	1.401(11)	C9–C10–C11	120.5(6)
C15–C16	1.375(12)	C10–C11–C12	120.5(8)
C16–C17	1.368(12)	C11–C12–C7	119.4(8)
C17–C18	1.394(10)	N3–C13–C14	118.9(5)
C18–C13	1.400(9)	N3–C13–C18	121.6(5)
		C14–C13–C18	119.4(6)
		C13–C14–C15	120.4(7)
		C14–C15–C16	120.2(8)
		C15–C16–C17	119.6(8)
		C16–C17–C18	121.2(7)
		C17–C18–C13	119.2(6)

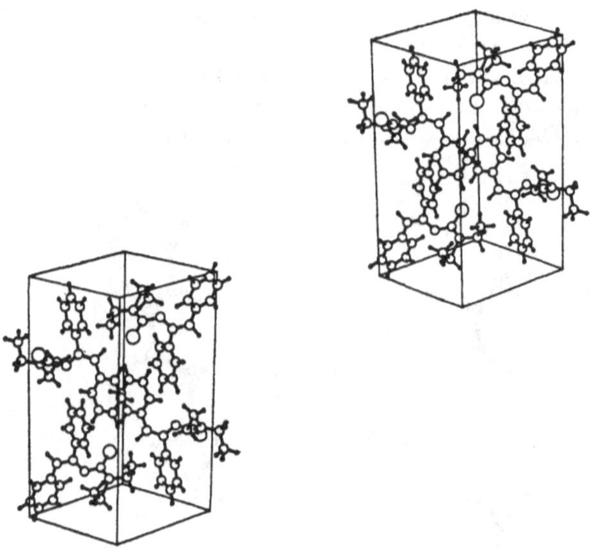

Fig. 3. Stereoscopic view of the molecular packing (MOTHERWELL, CLEGG)

Within the S,C2,N1,C1,N3 unit there are two types of C–N bonds. While the C1–N1 bond (1.289 Å) is a double bond the C2–N1 bond (1.382 Å) has a value halfway between single and double bond. The C2–N2 bond as well as the C1–N3 bond has a value falling in this range, too.

In comparison to these C–N bonds the N3–C13 bond (1.410 Å) is somewhat elongated.

Within the diethylamino moiety two bond lengths deviate extremly from normal values. Thus N2–C3 (1.74 Å) is much longer than a C–N single bond (1.47 Å) whereas C3–C4 (1.31 Å) is essentially shorter than the corresponding C–C single bond (1.54 Å). These unusual bond lengths which are accompanied by large B_{iso} values of the corresponding atoms are caused by strong thermal vibrations of this ethyl group.

Both phenyl rings are coplanar within experimental error making a torsion angle of 83.5° to each other.

Figure 3 illustrates the crystal structure of the title compound. The molecules are connected by N–H ... S interactions forming infinite chains. The S ... H distance (2.46 Å) is significantly shorter than the sum of the van der Waals radii. The angle N3–H1 ... S is 161°.

We thank Prof. Dr. L. Beyer for leaving the crystal material.

References

AHMED, F. R., HALL, S. R., PIPPY, M. E., SAUNDERSON, C. P.: NRC Crystallographic Computer Programs for the IBM/360 system. National Research Council, Ottawa 1966

BEYER, L., HARTUNG, J., WIDERA, R.: Tetrahedron **40**, 405 (1984)

BRAUN, U., RICHTER, R., SIELER, J., BEYER, L., LINDQVIST, O., YANOVSKY, A. I., STRUCHKOV, Yu. T.: Acta Cryst. **C43**, 92 (1987) a

BRAUN, U., RICHTER, R., SIELER, J., BEYER, L., LEBAN, I., GOLIČ, L.: Acta Cryst., publication submitted (1987) b

DIAS, H. W., TRUTER, M. R.: Acta Cryst. **17**, 937 (1964)

ELCOMBE, M. M., TAYLOR, J. C.: Acta Cryst. **A24**, 410 (1968)

JOHNSON, C. K.: ORTEP. Report ORNL-3794. Oak Ridge National Laboratory, Tennessee 1965

MOTHERWELL, W. D. S., CLEGG, W.: PLUTO 78. Program for Platting Molecular and Crystal Structures. Univ. of Cambridge, England (1978)

PAULING, L.: Die Natur der chemischen Bindung, pp. 214, Weinheim 1968

RICHTER, R., BEYER, L., ANDRIANOV, V. G., STRUCHKOV, Yu. T.: Z. anorg. allg. Chem. **518**, 123 (1984)

SHELDRICK, G. M.: SHELX 76, Program for Crystal Structure Determination. Univ. of Cambridge, England (1976)

SIELER, J., RICHTER, R., BEYER, L., LINDQVIST, O., ANDERSEN, L.: Z. anorg. allg. Chem. **515**, 41 (1984)

TRUTER, M. R.: J. Chem. Soc. 997 (1960) b

TRUTER, M. R.: Acta Cryst. **22**, 556 (1967) a

(Received July 15, 1987

Authors' address:

c/o Dr. Uwe BRAUN
Sektion Chemie der Karl-Marx-Universität Leipzig
Talstraße 35
Leipzig
DDR-7010

Cryst. Res. Technol.	23	1988	1	41–49

K. JURKSCHAT, G. WAGNER, P. PAUFLER

Wissenschaftsbereich Kristallographie der Sektion Chemie der
Karl-Marx-Universität Leipzig

Slip Band Formation during Bending of GaP Wafers

Dedicated to Professor Hermann NEELS on the occasion of his 75th birthday

{001} LEC-GaP:S wafers were deformed between 470 °C and 640 °C. {111} slip planes of maximum resolved shear stress were activated. Heights and densities of slip steps have been measured by optical and electron microscopy. At lower temperatures deformation proceeds mainly by growth of step heights whereas at higher temperatures bending is accomplished by the increase of slip band density rather then the height.

Blättchen von LEC-GaP:S mit 001-Orientierung wurden zwischen 470 °C und 640 °C verformt. Dabei wurden 111-Gleitebenen maximaler Schubspannung betätigt. Die Messung von Gleitstufenhöhen und -dichten erfolgte licht- und elektronenmikroskopisch. Bei tieferen Temperaturen schreitet die Verformung über das Wachstum von Stufenhöhen fort, während die Biegung bei höheren Temperaturen eher durch Zunahme der Gleitbanddichte realisiert wird.

1. Introduction

Plastic slip occurs by the generation and propagation of dislocations. A powerful tool for determining the dislocation processes during plastic deformation and the dislocation arrangement in deformed crystals is the study of slip lines and bands by optical and replica electron microscope techniques (cf., e.g., BLEWITT et al.; MADER). Under suitable conditions, averages over slip line heights, densities and lengths taken on the surface enable parameters of the bulk dislocation behaviour to be evaluated (cf., e.g., BERNER, KRONMÜLLER; NEUHÄUSER).

Evidence for slip lines in GaP has been obtained by several authors (BOEHNKE; DAVIDSON et al.; GERSHENZON, MIKULYAK; PAUFLER et al.). Slip markings always indicated slip along {111} planes and parallel to $1/2 \langle 1\bar{1}0 \rangle$ Burgers vectors which is in accordance with the zincblende structure ($F\bar{4}3m$) of that compound. Slip line parameters as mentioned above were evaluated for the first time by JURKSCHAT.

The aim of this communication is to report on a quantitative study of the geometry of slip in GaP wafers bent at elevated temperatures.

2. Experimental procedure

2.1. Specimen preparation

The work described here was performed using {001} LEC-GaP wafers 0.29 or 0.32 mm thick doped with sulphur to give a free carrier concentration of $(2 \ldots 6) \times 10^{17}$ cm^{-3}. The (001) surface of the wafers had been chemo-mechanically polished prior to plastic defor-

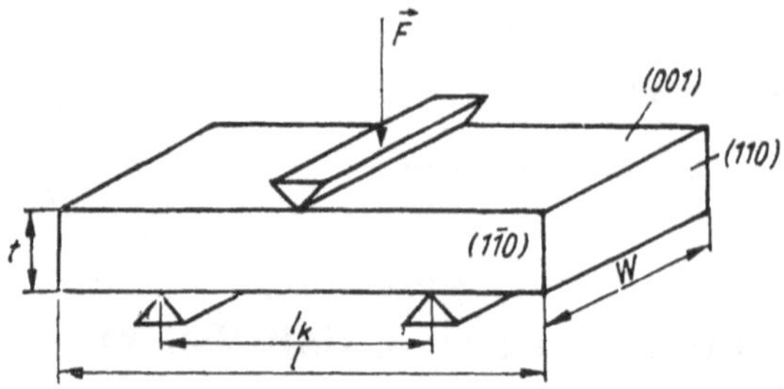

Fig. 1. Geometry of bending experiments. The axis of bending was [1$\bar{1}$0]

Table 1
Geometry of specimens (t thickness, w width, l length), etch pit density (EPD) and density of microdefects (ϱ_m) before deformation

Specimen No.	t mm	w mm	l mm
B 1	0.29	9.3	15.1
B 5	0.29	8.4	16.5
B 6	0.29	9.1	14.2
B 7	0.29	8.9	17.9
B14	0.32	9.8	15.5
B15	0.32	9.8	16.0
B16	0.32	9.1	15.6
B17	0.32	8.5	13.0

Specimens No. B 1 ... B 7: EPD $= 2.5 \times 10^4 \ldots 2 \times 10^5\ \text{cm}^{-2}$
$\varrho_m = (1 \ldots 2) \times 10^7\ \text{cm}^{-2}$
Specimens No. B 14 ... B17: EPD $= 2 \times 10^4 \ldots 5 \times 10^5\ \text{cm}^{-2}$
$\varrho_m = (7 \ldots 8) \times 10^5\ \text{cm}^{-2}$

mation. The back sides were grinded only. The geometry of specimens is shown in Figure 1 and in Table 1.

In order to determine the etch pit density, a H_3PO_4 etchant has been used at 180–190 °C. The time of etching varied between 4 and 5 minutes. According to GOTTSCHALCH et al. dislocations intersecting the surface and microdefects are revealed by this procedure (see Tab. 1). Etch pit distribution and densities had been determined at an adjoining wafer to avoid etching of deformation specimens prior to deformation. Etch pit densities (EPD) are given in Table 1.

2.2. Deformation experiments

Specimens were deformed in a N_2/H_2 atmosphere by means of a three-point bending jig attached to a Polanyi machine (see Fig. 1). Graphite knife-edges were used (radii 0.5 mm and 10 mm for the outer and inner knifes, respectively). The distance of the outer knife-edges was $l_k = 10.8$ mm. Force F and deflection d of the central knife have been measured.

The cross head speed was kept constantly at 6 mm/h. Details of the deformation equipment are described elsewhere (RETZLAFF).

The range of deformation temperature, T_{def}, as determined with Pt–PtRh thermocouples was 470 ... 640 °C (cf. Tab. 2). Specimens were slowly heated up to T_{def}. Deformation was started approximately 90 minutes after the beginning of heating. After that specimens were cooled down from T_{def} to 300 °C within about 20 minutes.

The bending angle α has been measured by an optical 2 circle goniometer. Variation of α across the wafer remained less than 0.2° except for specimens No. B 16 and B 17 (cf. Fig. 2). In the latter cases regions of nearly constant bending angle have been dealt with separately.

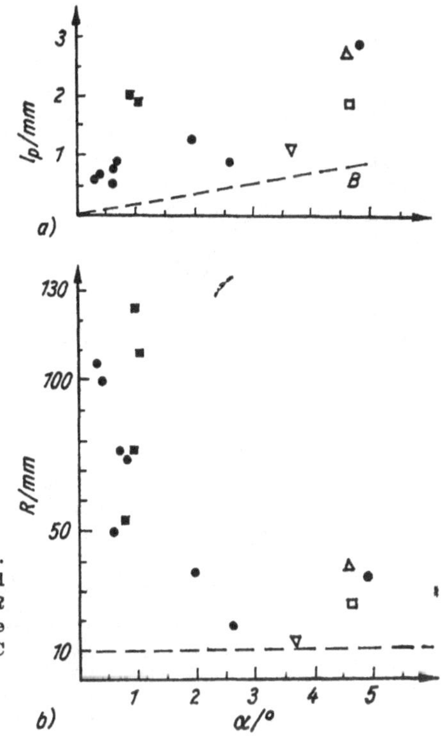

Fig. 2. (a) Length l_p of plastic zone versus bending angle α. The dashed line represents length of contact B calculated from the knife edge radius r; (b) Radius of curvature R versus bending angle α. The dashed line corresponds to the minimum radius $R_{min} = r$. T: 600 °C (●) 550 °C (△), 500 °C (□), 640 °C (▼), 470 °C (■)

From simple geometry it follows that the radius of curvature, R, depends on the length of the plastic zone l_p, the angle α and the wafer thickness t according to

$$R = t/2 + \frac{l_p}{2 \sin \alpha/2} \tag{1}$$

provided a sharp boundary between plastic and elastic zone. l_p has been determined by dislocation etching (cf. Fig. 3).

2.3. Microscopic techniques

Distances between and heights of slip steps have been measured by an optical microscope using the interference technique. Both signs of slip steps were observed but not analyzed separately. The height of slip steps was also estimated by a two step replica technique (Tylose + C/Pt). The angle between shadowing direction and surface was approximately 50°. The replicas have been investigated in a transmission electron microscope operating at 90 kV accelerating voltage.

3. Results

After deformation the etch pit density (EPD) increased considerably within a region defined by the length of the plastic zone l_p. Figure 3 gives an example. EPD's outside the plastic zone do not differ essentially from the values before deformation. The length l_p as determined by dislocation etching exceeds the length of contact B between knife and specimen perpendicular to the bending axis taken from

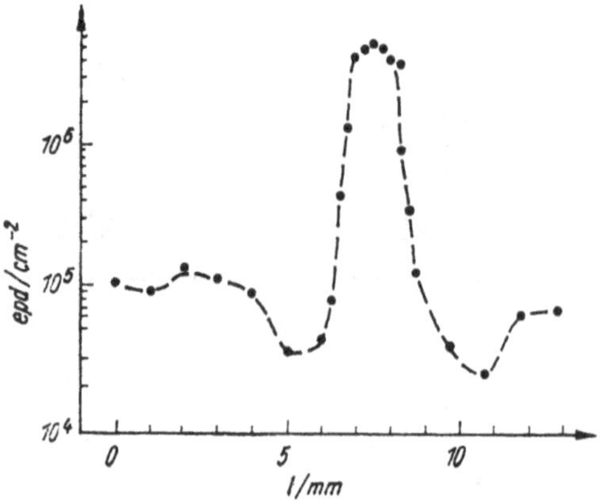

Fig. 3. Etch pit density (EPD) after bending measured parallel to the long edge of the specimen No. B 5

Table 2
Deformation Parameters (T_{def} temperature, α angle of bending, l_p length of plastic zone, R radius of curvature, $F^{(u)}$ upper yield force, $d^{(u)}$, $d^{(l)}$ deflection at upper and lower yield, respectively)

Specimen No.	T_{def} °C	α °	l_p mm	R mm	$F^{(u)}$ N	$F^{(l)}$ N	$d^{(u)}$ μm	$d^{(l)}$ μm	$(EPD)_{max}$ cm^{-2}
B 1	600	2.0	1.25	36.9	1.56	—	23	—	$1.1 \cdot 10^7$
B 5	600	4.9	2.9	34.4	7.08	6.65	12	33	$5.3 \cdot 10^6$
B 6	550	4.7	2.7	36.7	7.22	6.52	8	70	$6.6 \cdot 10^6$
B 7	500	4.7	1.9	23.3	8.3	6.79	10	48	$1.4 \cdot 10^7$
B 14	600	2.6	0.9	19.4	3.99	2.52	130	225	$8.8 \cdot 10^6$
B 15	640	3.6	1.1	14.4	4.4	3.45	104	240	$6.7 \cdot 10^6$
B 16/1	600	0.3	0.6	108.3	5.6	—	265	—	$3.4 \cdot 10^5$
B 16/2	600	0.4	0.7	100.5	5.6	—	265	—	$9.8 \cdot 10^5$
B 16/3	600	0.6	0.5	50.5	5.6	—	265	—	$5.1 \cdot 10^6$
B 16/4	600	0.7	0.9	77.2	5.6	—	265	—	$3.8 \cdot 10^6$
B 16/5	600	0.6	0.8	74.4	5.6	—	265	—	$6.6 \cdot 10^6$
B 17/1	470	0.7	—	—	6.5	—	315	—	$1.4 \cdot 10^5$
B 17/2	470	0.8	0.7	6.53	6.5	—	315	—	$1.2 \cdot 10^6$
B 17/3	470	0.9	1.2	76.5	6.5	—	315	—	$3.3 \cdot 10^6$
B 17/4	470	0.9	2.0	124.7	6.5	—	315	—	$3.6 \cdot 10^6$
B 17/5	470	1.0	1.9	109.0	6.5	—	315	—	$3.8 \cdot 10^6$

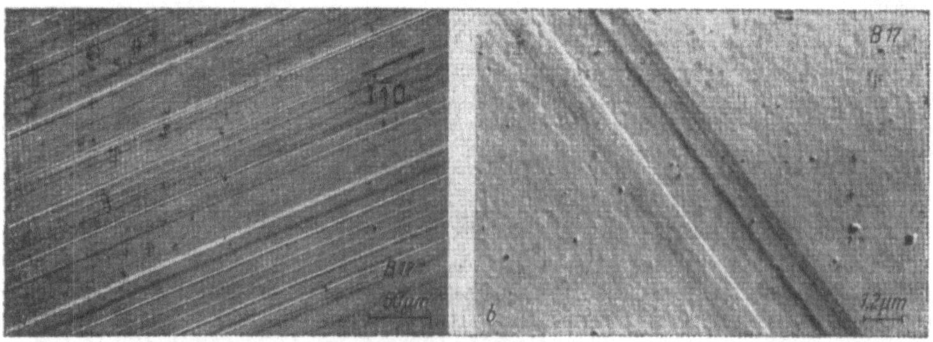

Fig. 4. Slip bands on (001) GaP (specimen No. B 17) bent at 470 °C. Angle of bending $\alpha \approx 1°$. (a) interference contrast technique using optical microscopy; (b) C/Pt replica studied by electron microscopy (specimen No. B 17), glide lines parallel bending axis [1$\bar{1}$0]

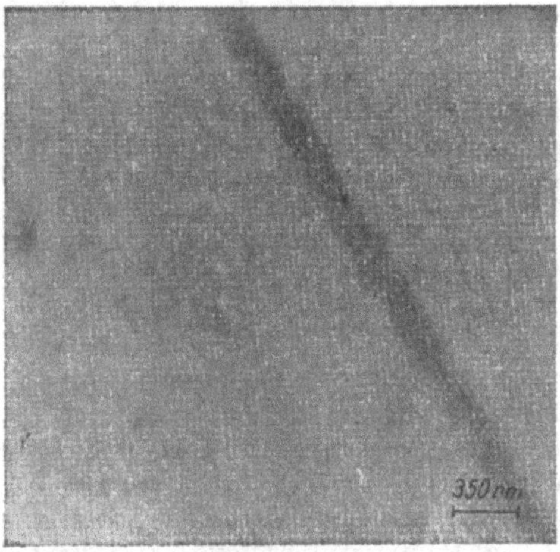

Fig. 5

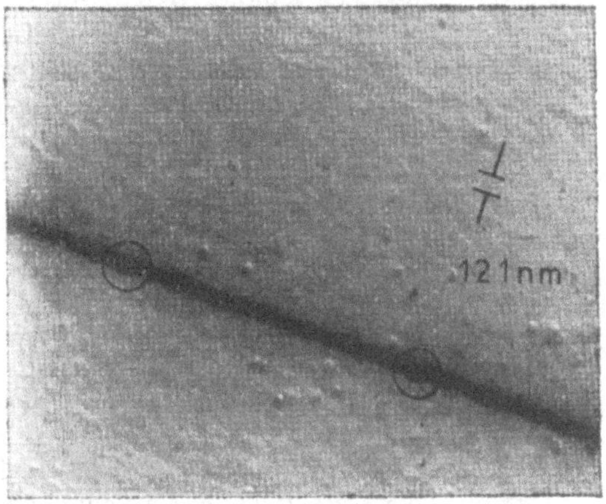

Fig. 6

Fig. 5. Slip lines clustered to form a slip band (specimen No. B 6, T_{def} 550 °C, α = 4.7 °)
Fig. 6. Slip step jogged twice as indicated by circles (specimen No. B 1, T_{def} = 600 °C, α = 2 °)

Table 3
Slip step parameters after bending of (001) GaP wafers at T_{def} (H_{max}^{LM}, H_{max}^{TEM} maximum heights as observed by optical and electron microscopy, respectively; $\bar{H}^{TEM}$ average heights; $\bar{A}$ average distances determined by optical microscopy)

Specimen No.	T_{def} (°C)	H_{max}^{LM} (nm)	H_{max}^{TEM} (nm)	$\bar{H}^{TEM}$ (nm)	$\bar{A}$ (μm)
B 1	600	179	178	82	9.2
B 5	600	—	—	—	5.9
B 6	550	137	289	84	2.9
B 7	500	137	295	116.5	2.7
B14	600	126			
B16/1	600	78			23.8
B16/2	600	113			14.75
B16/3	600	125			10.1
B16/4	600	107			7.0
B16/5	600	94			9.3
B17/2	470	138	166	70	10.2
B17/3	470	138	235	67	10.2
B17/4	470	138	337	128	10.2
B17/5	470	138	327	140	10.2

$B = 2r \sin(\alpha/2)$, where r = 10 mm is the radius of knife edge (cf. Fig. 2a). Note that l_p values which deviate most from the $B(\alpha)$ curve have been obtained with specimens of high upper yield point $F^{(u)}$ (see Tab. 2). In those cases dislocation movement has also taken place far from the region of direct contact. A similar conclusion has been drawn by Höche and Schreiber. The effective radius of curvature, R, decreases with increasing angle of bending until $R \approx r$ (cf. Fig. 2b). The increase of R at $\alpha > 4°$ is fictitious because the corresponding stress level again exceeded that of the other points considerably.

In Figure 4 typical slip band structures on (001) compression sides of samples are shown. Most of the slip steps higher than 100 nm appear as clusters of slip lines

(Fig. 5). Occasionally jogged slip steps were observed (Fig. 6). Results of measurement of slip step heights H and distances A are summarized in Table 3. Between relatively sharp slip bands of heights $>$ 100 nm and at average distances of (20 ... 30) μm apart numerous faint surface steps (heights between 5 and 50 nm) have been observed which were too close together in order the distances between them to be resolved by light microscopy. In this case the terminus "glide band" as defined by NEUHÄUSER should apply. Glide bands are rarely found in the transition region between plastic and elastic zones.

4. Discussion

Dislocations propagate through the crystal during bending along the {111} planes of maximum resolved shear stress (111) and ($\bar{1}\bar{1}1$). Different slip planes produce slip steps that appear dark and light, respectively (cf. Fig. 4).

From the results of Table 3 we conclude that at temperatures above 500 °C bending is accomplished mainly by a decrease of the average slip band distance $\bar{A}$ at nearly constant heights of the slip steps H (cf. Figs 7 and 8). On the other hand specimen No. B 17 deformed at 470 °C exhibited constant distances and step heights which increase with increasing deformation. The average number of dislocations $\bar{N}$ producing one slip step is estimated from H by

$$\bar{N} = \frac{\bar{H}}{\boldsymbol{b} \cdot [001]/a} = \frac{2\bar{H}}{a} \tag{2}$$

(see Fig. 9), where $\boldsymbol{b} = 1/2\,[\bar{1}01]$ is the Burgers vector of gliding dislocations and $a = 0.54515$ nm is the lattice parameter. Results are summarized in Table 4. In

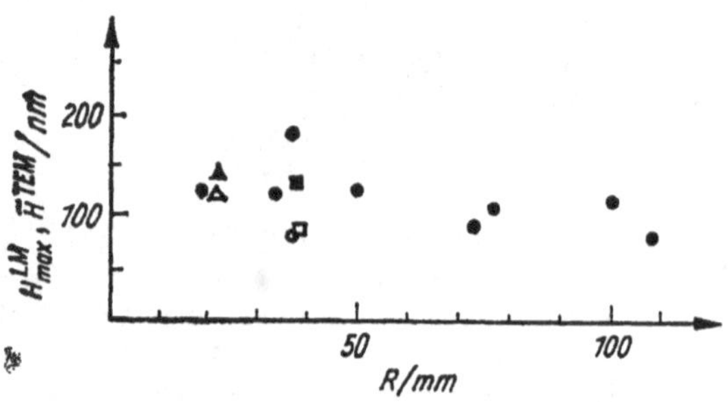

Fig. 7

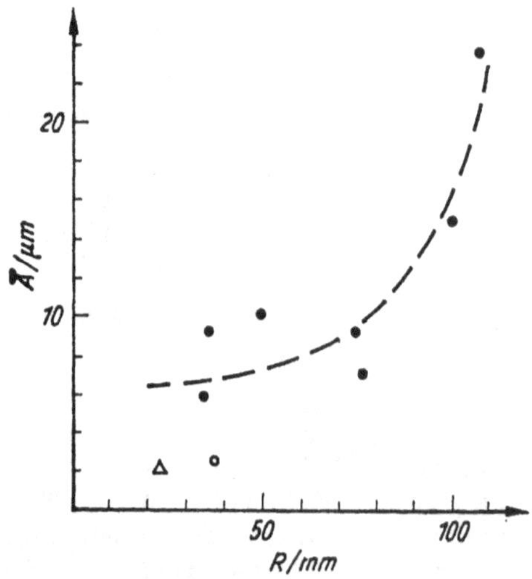

Fig. 8

Fig. 7. Heights H of slip bands versus radius of curvature R. ● LM, ○ TEM (T = 600 °C); ■ LM, □ TEM (T = 550 °C); ▲ LM, △ TEM (T = 500 °C)

Fig. 8. Average slip band distance $\bar{A}$ versus radius of curvature R. ● T = 600 °C, ○ T = 550 °C △ T = 500 °C

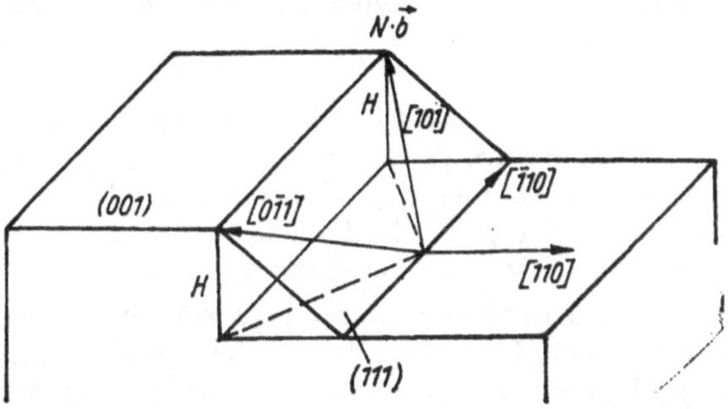

Fig. 9. On the relation between slip step height H and number N of dislocations producing the offset on a (001) surface by a shear process parallel to the Burgers vector $b = 1/2[\bar{1}01]$ and the slip plane (111). Each perfect dislocation carries a unit displacement of b

Table 4
Number of dislocations N per slip band according to eq. (2) and average shear strain $\bar{\gamma}$ (eq. (3))
N_{max} is estimated from $H = H_{max}^{TEM}$, $\bar{N}$ from $\bar{H} = \bar{H}^{TEM}$ (cf. Tab. 3)

Specimen No.	T_{def} (°C)	N_{max}	$\bar{N}$	$\bar{\gamma}$ (10^{-3})
B 1	600	653	301	13.4
B 6	550	1060	308	43.4
B 7	500	1082	427	64.7
B16/1	600	286*)		
B16/2	600	415*)		
B16/3	600	459*)		
B16/4	600	393*)		
B16/5	600	345*)		
B17/2	470	609	257	10.4
B17/3	470	862	240	9.9
B17/4	470	1236	470	18.8
B17/5	470	1200	514	20.6

*) $H = H_{max}^{LM}$

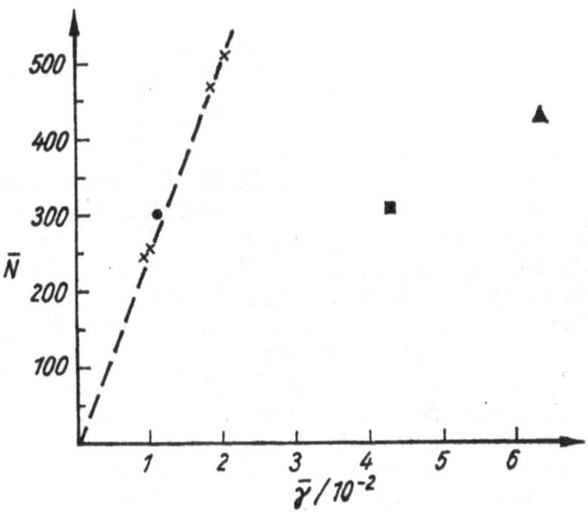

Fig. 10. Average number of dislocations $\bar{N}$ per slip band versus average shear strain $\bar{\gamma} = 3\bar{H}/2\bar{A}$ (cf. Tabs. 3 and 4). × T = 470 °C, ▲ 500 °C, ■ 550 °C, ● 600 °C

addition maximum values N_{max} and the average shear strain

$$\bar{\gamma} = \bar{H}/(\bar{A} \sin^2 \delta) = 3\bar{H}/2\bar{A} \tag{3}$$

are given where $\delta = 54.73°$ is the angle between slip plane and (001) surface.

Since there are different types of dislocations active during {111} slip (cf. Jurkschat; Paufler et al.). N gives some lower limit of the total number needed to produce a step of height H. Due to the limited resolution of the slip band observation technique N will not necessarily be the number per single slip plane.

At lower deformation temperature (specimen No. B 17, $T_{def} = 470$ °C) $\bar{N} \sim \bar{\gamma}$ has been observed (cf. Fig. 10). If the total length of dislocation lines per unit volume is L_v/V, the area δA swept out during an increment of average strain $\bar{\gamma}$ is $\alpha L_v \delta L_v$ so that

$$\bar{\gamma} = \alpha \bar{N} b L_v \delta L_v / V \tag{4}$$

where α is a geometric factor and b Burgers vector modulus.

From Figure 10 we estimate $\alpha L_v \delta L_v b/V \approx 4 \times 10^5$ and $\alpha L_v \delta L_v/V \approx 1 \times 10^{-4}$ $(nm)^{-1}$. The volume V between adjacent slip bands is $V = \bar{A}tw$ so that for specimen B 17 we find $\alpha L_v \delta L_v/A \approx 0.27$ m and $\alpha L_v \delta L_v \approx 2.8 \times 10^{-6}$ m^2. Dislocations move over distances of the order $\sqrt{\alpha L_v \delta L_v} \approx 10^{-3}$ m.

This agrees with the observation that slip steps extend over the whole specimen.

As will be shown in a subsequent paper dislocations are generated mainly at the surface during the first stage of deformation. Microdefects may also act as dislocation sources (Jurkschat).

5. Conclusion

From the present investigation we conclude that at relatively low temperatures (< 500 °C) deformation is more constricted to few slip bands and proceeds mainly by the growth of step heights and at elevated temperatures rather by the increase of slip band density. On the average 300–500 dislocations per band are moved whereas maximum numbers observed are of the order of 1000.

The authors would like to acknowledge the continuing support of experimental work by Mrs. B. Wegener and Mr. W. Wetzel.

References

Berner, Kronmüller

Blewitt, T. H., Coltman, R. R., Redman, J. K.: Report Conf. on Defects in Crystalline Solids, London: The Physical Society 1955, p. 369

Boehnke, O.: Diploma work, Karl-Marx-Universität Leipzig, Sektion Chemie 1981

Davidson, S. M., Iqbal, M. Z., Northrop, D. C.: phys. stat. sol. (a) **29**, 571 (1975)

Gershenzon, M., Mikulyak, R. M.: J. Appl. Physics **35**, 2132 (1964)

Gottschalch, V., Heinig, W., Butter, E., Rosin, E., Freydank, G.: Cryst. Res. Technol. **14**, 563 (1979)

Höche, H. R., Schreiber, J.: phys. stat. sol. (a) **86**, 229 (1984)

Jurkschat, K.: Diploma work, Karl-Marx-Universität Leipzig, Sektion Chemie, 1986

Mader, S.: Z. Physik **149**, 73 (1957)

Neuhäuser, H.: in: Dislocations in Solids, Vol. 6, ed. by F. R. N. Nabarro, Amsterdam 1983, pp. 319

PAUFLER, P., ROTSCH, P., WAGNER, G.: Philos. Mag. (1987)
RETZLAFF, U.: Diploma work, Karl-Marx-Universität Leipzig, Sektion Chemie 1985
ROTSCH, P.: Diploma work, Karl-Marx-Universität Leipzig, Sektion Chemie 1985

(Received July 15, 1987)

Authors' address:

Dipl.-Krist. Kerstin JURKSCHAT, Dr. rer. nat. Gerald WAGNER,
Prof. Dr. sc. nat. Peter PAUFLER
Karl-Marx-Universität Leipzig, Sektion Chemie, Wissenschaftsbereich Kristallographie
Scharnhorststr. 20
Leipzig
DDR-7030

Cryst. Res. Technol.	23	1988	1	51–57

K. Fichtner

Zentralinstitut für physikalische Chemie der Akademie der Wissenschaften der DDR

OD Approach to the Polytypism in CdP_2

Dedicated to Professor Hermann Neels on the occasion of his 75th birthday

The structures of CdP_2 are described as stacking sequences of layers of CdP_4 tetrahedra. On the basis of the OD theory of Dornberger-Schiff, the stacking possibilities are derived and convenient stacking notations proposed. The idealized layer structure is given, the percentage of α-CdP_2 for any stacking sequence defined. The symbol of the OD groupoid family is P11(n)2m | (4) x, y with parameters $x = y = 1/2$ and (4^+), (4^-) as the two alternative transformations from layer to layer. The net constants are $a = b \approx 5.29$ Å; the "thickness" of a layer is $c_0 \approx 4.95$ Å.

Die Strukturen von CdP_2 werden als Stapelfolgen von Schichten von CdP_4-Tetraedern beschrieben. Auf der Grundlage der OD-Theorie von Dornberger-Schiff werden die Stapelmöglichkeiten abgeleitet und zweckmäßige Polytypensymbole vorgeschlagen. Die idealisierte Schichtstruktur wird angegeben und der Prozentsatz von α-CdP_2 in einer beliebigen Stapelfolge definiert. Das Symbol für die OD-Gruppoid-Familie ist P11(n) 2m | (4 x, y) mit Parametern $x = y = 1/2$ und (4^+), (4^-) als den zwei möglichen Transformationen von Schicht zu Schicht. Die Netzkonstanten sind $a = b \approx 5.29$ Å, die „Schichtdicke" ist $c_0 \approx 4.95$ Å.

1. Introduction

Cadmium diphosphide exists in a low-temperature orthorhombic α-form (Goodyear, Steigmann; Olofsson, Gullman) and in a high-temperature tetragonal β-form (Stackelberg; Paulus; Parthé). There is a very close relationship between the lattice parameters of both forms, and the volume of the unit cell of the β-phase is twice that of the α-phase (compare Tab. 1). Manolikas et al. have shown that the two polymorphs may be understood as different stackings of the same layer-like structural building units, i.e. CdP_2 is a polytypic compound. The intermediate phases in the phase transition from α- to β-CdP_2 are supposed to be polytypes, whose stacking sequences are mixtures of α and β. The percentage of β-type stacking rises with temperature.

The theory of order-disorder structures (OD theory) of Dornberger-Schiff (1956, 1979, 1982) explains the existence of different stacking variants of a layered structure by partial symmetry operations, which are valid for any single layer but do not extend to its adjacent layers. The different structures are built up of the same kind of layers and the same pairs of layers; they differ, however, in triples and higher n-triples of layers.

The aim of this paper is to apply the OD theory to CdP_2 in order to reach a better understanding and a convenient description of the polytypism in this compound.

Table 1
Crystal data of the CdP_2 phases

phase	space group	lattice parameters (Å)	reference
α	$P\,na2_1$	$a = 9.90$, $b = 5.408$, $c = 5.171$	OLOFSSON, GULLMANN
β	$P\,4_12_12$	$a = 5.29$, $c = 19.74$	STACKELBERG, PAULUS

2. Choice of layer and family of OD structures

The two known structures of CdP_2 may be considered as consisting of planes of atoms (MANOLIKAS et al.). Perpendicular to the z-direction of the β-phase, which corresponds to the x-direction of the α-phase, there are planes of phosphorus atoms (P-planes) and planes consisting of cadmium as well as of phosphorus atoms ("mixed" or M-planes). In both observed structures, one P-plane alternates with two M-planes (Fig. 1).

For the description of the different stacking variants of a polytypic substance, it is of advantage to consider layerlike building units, which are as thick as possible. If we demand that the symmetry operations in the known structures transform whole layers into each other, there are two possible choices of the layer:

(i)

... —P—M—M—P—M—M—P—M—M—P— ...

(ii)

The "triple" layer in the first choice consists of three atomic planes: one P-plane and its two adjacent M-planes. This layer has been chosen by MANOLIKAS et al.

The "OD layer" or "layer of tetrahedra" in the second choice consists of two consecutive M-planes. Any P-plane is the boundary between two layers and may be considered as belonging to both of them. Thus, this layer consists of four atomic planes, P—M—M—P (Fig. 2). In CdP_2 any Cd atom is tetrahedrally coordinated by four P atoms, and any structure may be represented by corner-connected CdP_4 tetrahedra (Fig. 3). The OD layer consists of parallel chains of tetrahedra.

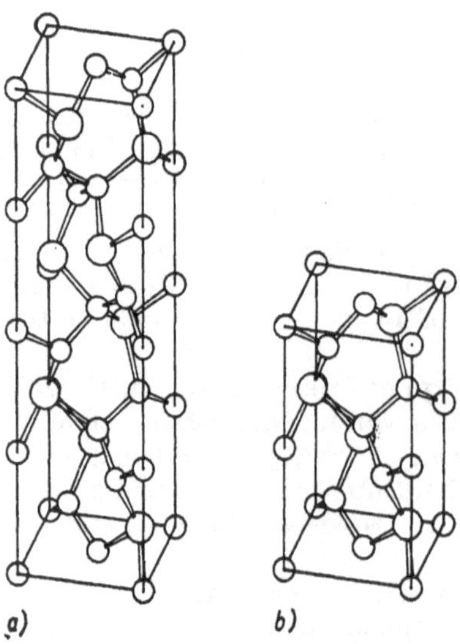

a) b)

Fig. 1. The two phase in Cadmium diphosphide. a) α-CdP_2, b) β-CdP_2

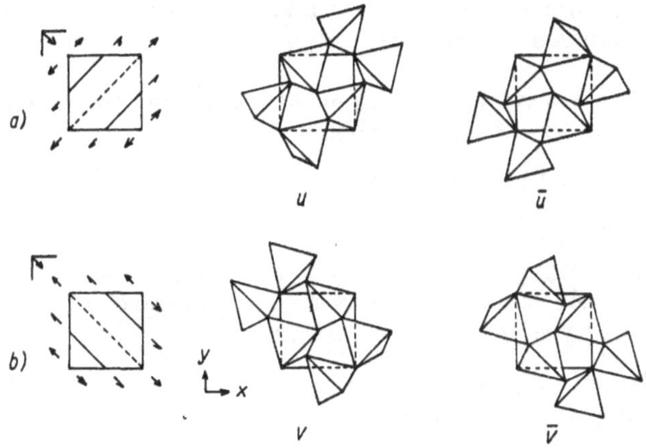

Fig. 2. Layer group diagrams and possible orientations of the OD layers in CdP_4. Any layer is represented by two chains of tetrahedra. a) layer group P 11(n)2m, orientations u and $\bar{u}$; b) layer group P 11(m) m2, orientations v and $\bar{v}$

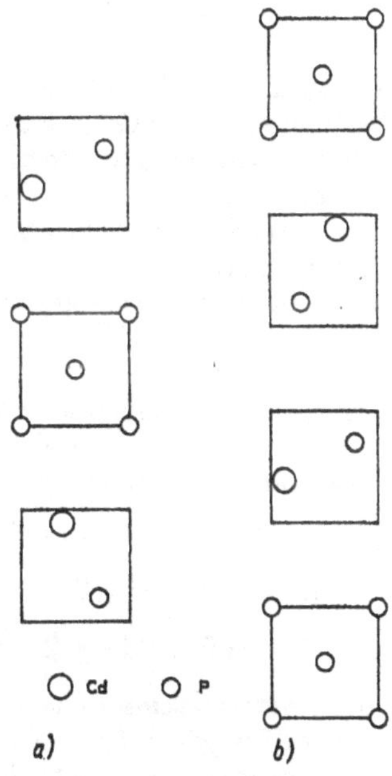

Fig. 3. Atomic planes for the two choices of layer .a) triple layer M-P-M; b) OD layer P-M-M-P

The layers of tetrahedra are appropriate for the application of the OD theory. The symmetry of the layer may be deduced from the symmetries present in the two crystal phases. In the β-phase, the layers are transformed into itself by 2-fold rotations around axes in the directions $x + y$ or $y - x$, alternating in the sequence of layers. The α-phase, referred to the axes of the β-phase, has the space group P 2_1cn. The n-glide reflection transforms the layer into itself. Thus, referred to directions x, y, z, $x + y$, $y - x$, the layer group symbol is P 11(n)m2 or P 11(n)2m, alternating in both crystal phases. These symbols correspond to the layer group symbols Cm2(a) and C2m(a), respectively. For both crystal phases, the reflection is only a partial symmetry. It does not hold for the adjacent layers. But this reflection may be used to explain the stacking possibilities. If a certain layer L_1 and one possible orientation L_2' of the next layer is given, the reflection of L_1 transforms L_2' into the alternative orientation L_2''.

Figure 3 shows the four possible orientations of the layers. Any OD layer is transformed into the next one by the following operations

$$\text{either} \quad c.(.),\ .2_1(.),\ (4^+) \text{ and } (\bar{4}^-)$$

$$\text{or} \quad .c(.),\ 2_1.(.),\ (4^-) \text{ and } (\bar{4}^+)\,.$$

The two sets of transformations correspond to the two possible orientations of the next layer. With the 4-fold screw rotation as the most characteristic transformation from layer to layer, the short symbol of the OD groupoid family becomes

$$P\,11(n)2m\,|(4)\ (x, y, z)$$

$$(4^+)\ (0.5, 0.5, 1)\ ;\ (4^-)\ (0.5, 0.5, 1)\,.$$

CdP_2 is the first substance known with this OD groupoid family.

3. The structure of a layer

The structure of the CdP_2 layer is derived from the structure reported for the α-phase by OLOFSSON and GULLMAN. For this, some idealization is necessary: the net has to be made quadratic, and the atomic positions have to be slightly shifted in order to fulfill the mirror plane of the OD layer with layer group P 11(n)2m. Over this, the axes are interchanged in order to have the z-direction perpendicular to the layer. A P atom is chosen as origin.

The lattice parameters of the layer are defined as follows:

$$a = b = \sqrt{b_\alpha \cdot c_\alpha} = 5.288\ \text{Å}\,, \quad c_0 = a_\alpha/2 = 4.95\ \text{Å}\,.$$

The structure has three independent atomic positions: Cd, P1, and P2. The coordinates for the α-phase and for the layer are given in Table 2. The values for the layer have been calculated in two steps:

(i) Change of axes and origin, symmetrization of the z-coordinate of P2.

$$x(\text{I}) = y_\alpha(\text{I}) - y(\text{P2}),\ y(\text{I}) = z_\alpha(\text{I}) - z_\alpha(\text{P2}),\ \text{I} = \text{Cd, P1, P2}\ ;$$

$$z(\text{P2}) = 0,\ z(\text{Cd}) = 2 \cdot x_\alpha(\text{Cd}),\ z(\text{P1}) = 2 \cdot x_\alpha(\text{P1})\,.$$

(ii) Symmetrization of x- and y-coordinates, keeping interatomic distances about the same as given by OLOFSSON and GULLMAN. Cd and P1 have been shifted to the plane $x - y = -0.5$. Several positions have been tested in order to reach a good agreement with the interatomic distances as reported for the α-phase (compare Tab. 3).

Table 2
Atomic coordinates for α-CdP_2 according to OLOFSSON, GULLMANN (left) and after idealization and change of axes and origin (right)

	x/a_α	y/b_α	z/c_α	x/a	y/b	z/c_0
Cd	0.1529	0.1011	0.2606	−0.125	0.375	0.306
P(1)	0.1185	0.4442	0.5957	0.215	0.715	0.237
P(2)	−0.0064	0.2693	−0.1036	0	0	0

Table 3
Interatomic distances in Å according to OLOFSSON, GULLMANN (left) and after idealization (right)

Cd−P(1)	2.56	2.57	Cd−P(2)	2.62	2.58	P(1)−P(2)	2.17	2.22
	2.56	2.56		2.57	2.58		2.20	2.22

4. Stacking possibilities and polytype notations

According to the symbol of the OD groupoid family derived in section 2, any OD layer is transformed into the next one by a 4-fold screw rotation with rotation angle either $+90°$ or $-90°$. Thus, the stacking sequence may be characterized by the sequence of signs of the rotation angles. The α-phase has the symbol $+-$, the β-phase is characterized by $++++$, and the enantiomorph of the β-phase with space group $P\,4_32_12$ has the sequence $----$. Obviously, there are four possible orientations of the OD layers (Fig. 3). They may be denoted by letters u, $\bar{u}$, v, $\bar{v}$, which refer to the four directions of the chains of tetrahedra, $x+y$, $-x-y$, $y-x$ and $x-y$, respectively. In any stacking sequence, letters u and $\bar{u}$ are followed by v or $\bar{v}$, and vice versa. The α-phase has the stacking sequence uv; the β-phase is characterized by $uv\bar{u}\bar{v}$.

Each of the two proposed stacking notations has its advantages. The relations between them are described by the following formulae:

$$uv = +,\ v\bar{u} = +,\ \overline{uv} = +,\ \bar{v}u = +\ ;$$

$$u\bar{v} = -,\ vu = -,\ \bar{u}v = -,\ \overline{uv} = -\ .$$

Any sequence of letters uniquely determines a sequence of signs. If a starting letter is given, the sequence of letters follows from the sequence of signs.

The stacking possibilities may also be described in terms of atomic planes. The P-planes are always the same, independent on the stacking. If an M-plane is fixed, the adjacent M-plane is uniquely determined, and the M-plane after the next P-plane has two possible orientations. The atomic positions of the 4 orientations of the OD layers are characterized in Figure 4. A stacking notation for sequences of atomic planes has been proposed by MANOLIKAS et al. The symbols contain letters A, b_1, b_1', b_2, b_2', c_1, c_1', c_2, c_2'. The α-phase has the sequence $b_1Ac_1c_1'Ab_1'$, the β-phase is described by $b_2'Ac_1c_1'Ab_1'b_1Ac_2'c_2Ab_2$. A transformation between the notation of MANOLIKAS et al. and the notations proposed in this paper is possible by the following formulae:

$$Ab_1'b_1 = u\ ,\quad Ab_2b_2' = \bar{u}\ ,\quad Ac_1c_1' = \bar{v}\ ,\quad Ac_2'c_2 = v\ .$$

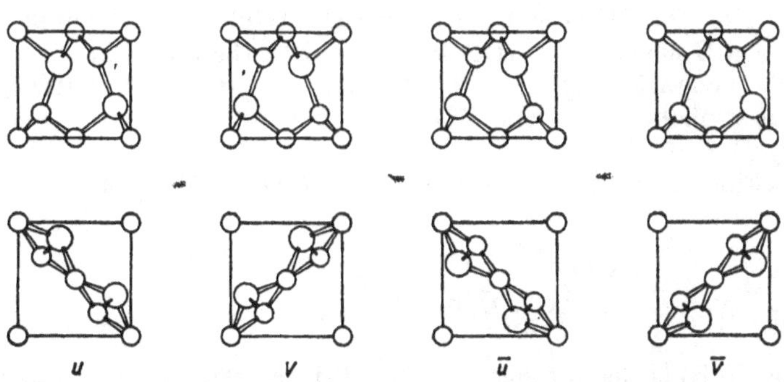

Fig. 4. The four orientations of the OD layer in projections along z (lower row) and along y (upper row)

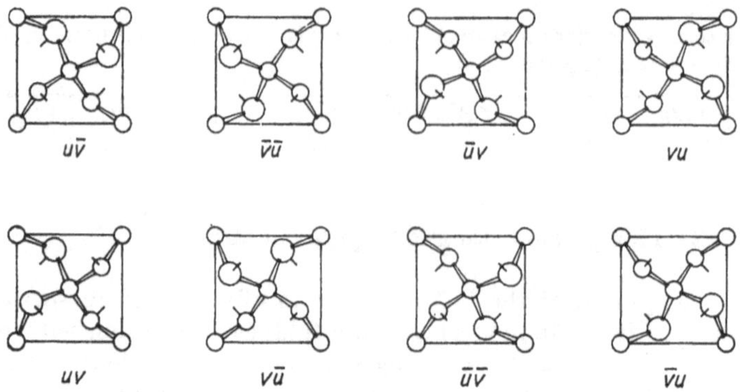

Fig. 5. The eight orientations of the triple layer, projection along z

The triple layer M—P—M, described in section 2, is representative for a pair of OD layers P—M—M—P—M—M—P, as adjacent M-planes uniquely determine each other and the P-planes are fixed. There are 8 possible orientations for the pairs of OD layers and thus for the triple layers (Fig. 5). Therefore, a characterization of the stacking by orientations of triple layers is less convenient than a characterization by orientations of OD layers. Any triple layer, however, may be given a sign + or −, as it is representative for a pair of OD layers. The sequence of signs defined for OD layers may be considered as a stacking notation for triple layers. Figure 5 contains the key on the possible successors of a certain triple layer. Example: $v\bar{u}$ may be followed by $\bar{u}\bar{v}$(= +) or $\bar{u}v$ (= −).

5. MDO structures and the percentage of α-phase

In OD theory, structures of maximum degree of order (MDO structures) are considered as the most simple and most probable structures. In order to determine the MDO structures for CdP_2, the triples of OD layers have to be considered.

There are two enantiomorphic pairs of OD layers, characterized by the signs + and −. Three non-congruent triples may be formed: $T_1 = ++$, $T_2 = --$, $T_3 = +- = -+$. Each of the triples uniquely determines a structure containing only this kind of triple. The MDO structure with the triple T_1 is denoted as MDO_1. It corresponds to the β-phase. The MDO structure with the triple T_2 is enantiomorphic to MDO_1, as T_2 is enantiomorphic to T_1. The triple T_3 determines MDO_2, which corresponds to the α-phase. Thus, the MDO structures corresponds to the two known phases of CdP_2.

For any stacking sequence, the percentage of MDO_2 (or α-phase) may be defined by

$$P = \frac{N(T_3)}{N(T_1) + N(T_2) + N(T_3)} \cdot 100\,.$$

In this formula, $N(T_i)$ is the number of triples T_i in a period of the stacking sequence. This number P may be useful for a correlation of structures and properties for intermediate phases in the transformation of α to β.

6. Concluding remarks

The application of OD theory to CdP_2 has given a clear description of the stacking possibilities and a number of useful tools for handling the polytypism in this compound. This can be of advantage in the further investigation of the intermediate phases in the phase transition from α- to β-CdP_2. Over this, the example of CdP_2 is also of interest to OD theory. No other structure with a similar stacking of layers has been found so far.

For ZnP_2, a tetragonal phase and a monoclinic phase have been observed, and the tetragonal phase is isotypic to the β-phase of CdP_2. It has to be proved, whether ZnP_2 may also be considered as a substance with OD character.

References

DORNBERGER-SCHIFF, K.: Acta Crystallogr. **9**, 593 (1956)
DORNBERGER-SCHIFF, K.: Kristall u. Technik **14**, 1027 (1979)
DORNBERGER-SCHIFF, K.: Acta Crystallogr. **A38**, 483 (1982)
GOODYEAR, J., STEIGMANN, G. A.: Acta Crystallogr. **B25**, 2371 (1969)
MANOLIKAS, C., VAN TENDELOO, J., AMELINCKX, S.: phys. stat. sol. (a) **97**, 87 (1986)
OLOFSSON, O., GULLMANN, J.: Acta Crystallorgr. **B26**, 1883 (1970)
PARTHÉ, E.: Crystal Chemistry and Tetrahedral Structures, New York 1964, p. 127
STACKELBERG, M., PAULUS, P.: Z. phys. Chem. **28B**, 427 (1935)

(Received July 22, 1987)

Author's address:

Dr. sc. K. FICHTNER
Zentralinstitut für physikalische Chemie der Akademie der Wissenschaften der DDR
Rudower Chaussee 5
Berlin
DDR-1199

Cryst. Res. Technol.	23	1988	1	59–67

G. WAGNER, V. GOTTSCHALCH

Karl-Marx-Universität, Sektion Chemie, Leipzig

Revealing of Lattice Defects on ($\bar{1}\bar{1}\bar{1}$) Faces of Gallium Phosphide and Indium Phosphide by Chemical Etching

Dedicated to Professor Hermann NEELS on the occasion of his 75th birthday

A solution is presented suitable for revealing lattice defects on ($\bar{1}\bar{1}\bar{1}$) phosphorous faces of GaP and InP. It is possible to distinguish between pits originated by grown-in dislocations and microdefects such as perfect loops, faulted loops and precipitates and/or inclusions. Moreover it is also possible to reveal large stacking faults of Shockley type and microtwin lamellae. One to one correlations have been given by means of transmission electron microscopy.

Es wird eine neue Lösung vorgestellt, die auf ($\bar{1}\bar{1}\bar{1}$) Flächen von GaP und InP Gitterdefekte anätzt. Es ist möglich zwischen Versetzungen und Mikrodefekten wie vollständigen Versetzungsringen, unvollständigen Versetzungsringen sowie Ausscheidungen und/oder Einschlüssen zu unterscheiden. Neben ausgedehnten Stapelfehlern vom Shockley-Typ werden auch Mikrozwillingslamellen und Korngrenzen angeätzt. Eine exakte Korrelation der verschiedenen Ätzerscheinungen mit den entsprechenden Gitterdefekten erfolgte mittels Transmissionselektronenmikroskopie.

1. Introduction

For characterization of substrates and epitaxial layers of GaP and InP optoelectronic devices the chemical etching proves a rapid and reliable method. Only two etchants have been mainly used for detailed investigations on ($\bar{1}\bar{1}\bar{1}$) phosphorous faces of GaP, the RC-etchant and AB-solution (RICHARDS, CROCKER; ABRAHAMS, BUIOCCHI). Not only the AB-solution but also the RC-etchant produce more or less well-defined dislocation etch pits (D-pits) and shallow pits (so-called S-pits or saucer pits) in which S-pits were attributed to microdefects. The etching behaviour of solution AB and RC-etchant had been extensively discussed elsewhere (RC-etchant: RICHARDS, CROCKER; BEPPU et al.; WERKHOVEN et al. 1979; TAJIMA; WERKHOVEN; WERKHOVEN et al. 1977; ROZGONYI et al.; STRINGFELLOW et al.; BRANTLEY et al.; ROZGONYI, IIZUKA; IIZUKA; AB-solution: ABRAHAMS, BUIOCCHI; STRINGFELLOW et al.; BRANTLEY et al.; ROZGONYI, IIZUKA; IIZUKA). Furthermore, other etchants were described for revealing dislocations and partly microdefects on ($\bar{1}\bar{1}\bar{1}$) phosphorus faces of InP grown by different methods (CLARKE et al.; KOMIYA, NAKAJIMA; KOTANI et al.; AKITA et al.; MULLIN et al.; CHU et al.). However, for revealing lattice defects on ($\bar{1}\bar{1}\bar{1}$) and (100) InP faces have been mostly used the AB-solution (ABRAHAMS, BUIOCCHI; KOMIYA, NAKAJIMA; KOTANI et al.; HUBER, LINH; BROWN et al. 1980) and H-etchant (HUBER, LINH; AKITA et al.; COCKAYNE et al. (a, b) 1981; BROWN et al. 1981).

The purpose of this paper is to describe the results obtained by a new etchant with respect to revealing grown-in dislocations, microdefects, stacking faults and microtwin lamellae on ($\bar{1}\bar{1}\bar{1}$) phosphorous faces of variously doped GaP and InP grown by different methods.

2. Experimental

The new etchant used consists of HNO_3 (conc.): H_2O_2 (30%) = 1:1 (vol/vol) (WAGNER, GOTTSCHALCH 1983). The etching procedure took place at boiling temperature and the etching times varied within the range of 20 seconds to 3 minutes. Material used for experiments has been grown by liquid encapsulation Czochralski (LEC) method, vapour phase epitaxy (VPE) and liquid phase epitaxy (LPE). From Table 1 it can be taken away characteristic data of materials used. The etch pattern were examined by optical interference contrast microscopy. GaP specimens for TEM investigations were thinned chemically using a solution consisting of HNO_3:HCl = 1:3 (vol/vol) at 323–333 K beginning at the (111) face. Before thinning procedure the GaAs substrates have been removed from the GaP heteroepitaxial samples by boiling nitric acid. After thinning the samples for the one to one correlation were treated with the new etchant mentioned above and examined in a transmission electron microscope operating at 90 kV accelerating voltage.

Table 1
Characteristical data of material used for experiments

material	growth method	growth conditions see	$N_D - N_A$ at 300 K (cm^{-3})	dopant
GaP	LEC		(3.2 ... 8.4) × 10^{17}	S + Si
GaP	VPE on GaAs	WAGNER et al.	(1.1 ... 3.4) × 10^{17}	oxygen
GaP	LPE on GaAs	GOTTSCHALCH et al., NEUMANN et al.	roughly 4 × 10^{17}	Sn
InP	LEC		5 × 10^{17} to 1 × 10^{18}	Sn

3. Results and discussion

3.1. VPE grown GaP containing oxygen

Due to the heteroepitaxial conditions and plastic deformation the VPE-grown GaP layers investigated contain a high number of dislocations, manyfold lying stacking faults of Shockley type and microtwin lamellae (WAGNER et al. 1978). Furthermore, during early growth stages an appreciable value of isolated twinned GaP hillocks is formed on the GaAs substrate. Consequently, subgrain twin boundaries (e.g. B and B_1 in Figure 2) are generated, but only directly at the GaAs/GaP interface (WAGNER, KUEHN). The etch pattern produced on the ($\bar{1}\bar{1}\bar{1}$) phosphorous surface of such layers consist of small (examples are arrowed in Figure 1) and large D-pits (D in Fig. 1) and two kinds of well-defined grooves. Ideally, the etch grooves bounded by two more or less distinct etch pits (some examples are labelled with D_P) are caused by stacking faults, whilst the grooves without bounding pits are attributed to microtwin lamellae. Pits bounding the etch grooves of stacking faults are developed by Shockley partials. Frequently, grooves consist of chains of etch pits in which the distance between single pits is extremely small (see SF_m in Fig. 1). Thus the etch pit chain is formed by either

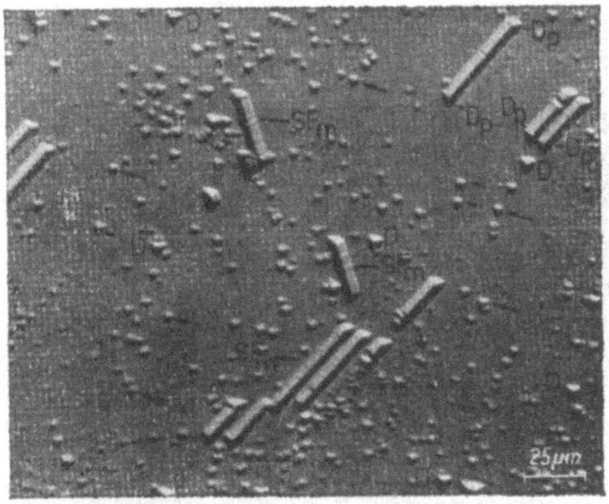

Fig. 1. Optical interference micrograph of etch pattern produced on the surface of a VPE grown GaP heteroepitaxial layer (thickness 200 μm) with orientation $(\bar{1}\bar{1}\bar{1})$; etching time 1.5 min; for explanation see text

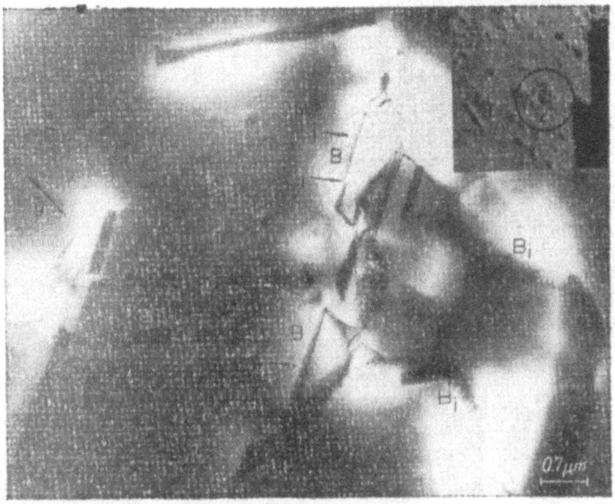

Fig. 2. Bright-field TEM image ($g = 220$) of a complex of subgrain twin boundaries near the GaAs/GaP interface corresponding to the etoh features (etching time 20 seconds) in the top right hand image corner; B_i – incoherent boundary; B – coherent boundary; j – jog of coherency plane

perfect dislocations interacting with coherent twin boundaries due to slip as a consequence of plastic deformation (SIETHOFF) (arrows in Fig. 3b) and/or partial dislocations originated by jogs of the {111} coherency planes (j in Fig. 2). Sometimes it is difficult to separate the various etch grooves mentioned. Both types of D-pits (see D and arrows in Fig. 1) are caused by threading dislocations. Namely, with increasing etching time the small D-pits arrowed in Figure 1 grow and do not disappear as e.g. the small etch pits caused by microdefects. Possibly the pit size is dependend on the dislocation type and/or intersecting angle between dislocation line and crystal surface (ROSIN, FREYDANK). The Figures 3a and b depict an exact one to one correlation between some etch features (Fig. 3a) and the true lattice defects observed by TEM (Fig. 3b). Due to the relatively short etching time of 20 seconds only stacking faults, microtwin lamellae and subgrain twin boundaries (see Fig. 2) have produced distinct etch features on thinned sample, however, the inclined dislocations no yet.

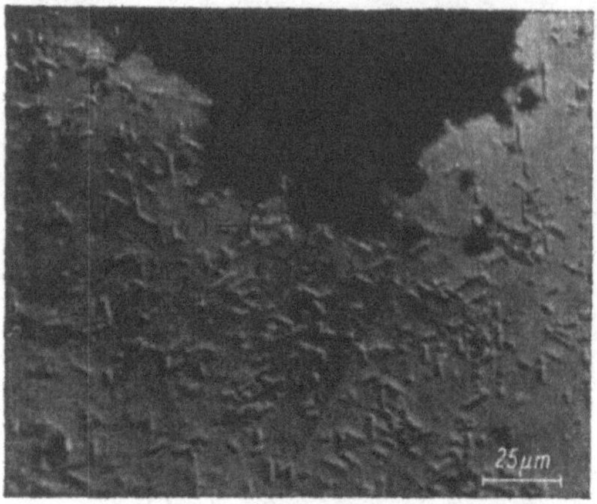

3a

Fig. 3. a) Optical interference micrograph of a thinned and afterwards etched sample (echting time 20 seconds); b) TEM micrograph (bright-field image with g = 220) of a defect arrangement corresponding to the encircled area; SF_m – manyfold overlapping stackingfaults; T – microtwin lamella

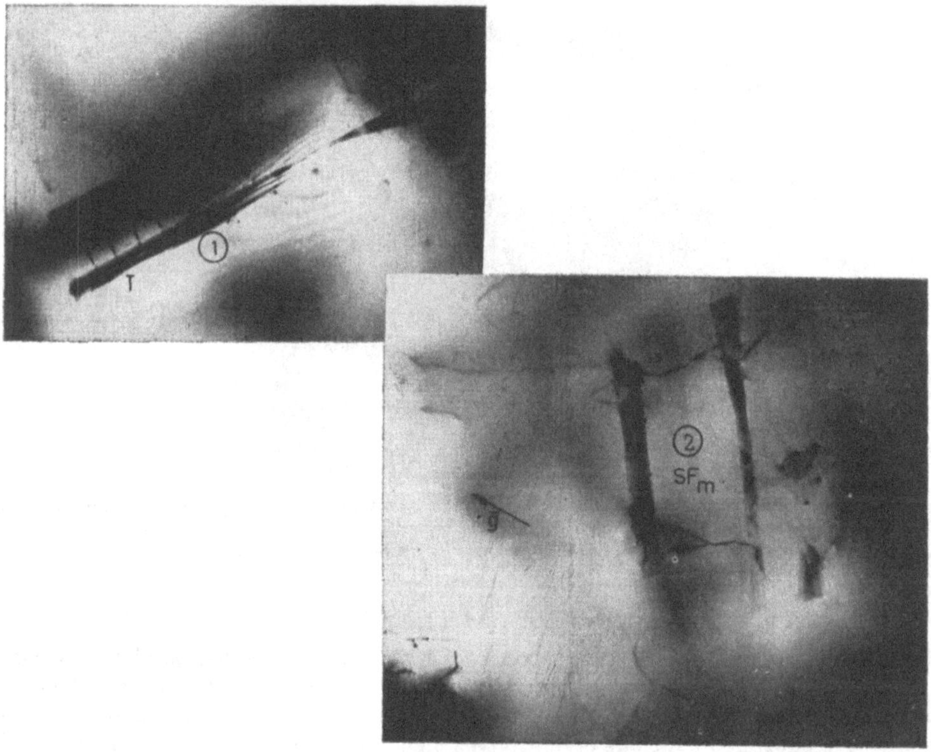

3b

3.2. LPE gallium phosphide grown from Sn solution

Generally on this material heavily doped with Sn and deposited onto GaAs could be revealed the same defects as on $(\bar{1}\bar{1}\bar{1})$ phosphorous faces of VPE grown layers. The grooves in Figure 4 correspond to manyfold lying stacking faults and microtwin lamellae. Additionally mutch small indistinct pits occur. They are attributed to helical dislocations (H in Fig. 5) and microdefects such as perfect dislocation loops (some examples are arrowed in Fig. 5) and faulted loops. Mostly such microdefects are

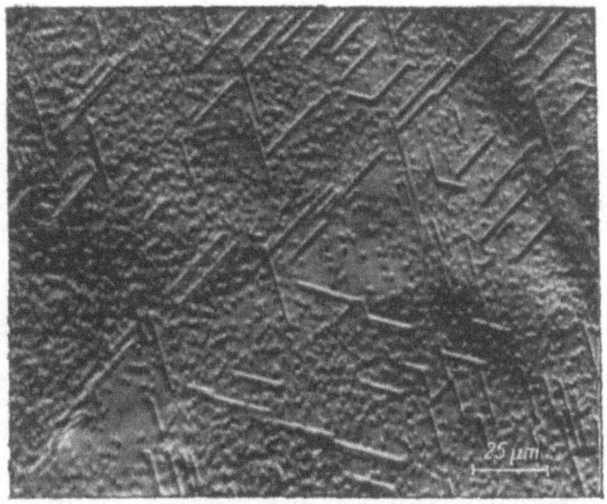

Fig. 4. Optical interference micrograph of etch pattern obtained on a $(\bar{1}\bar{1}\bar{1})$ oriented GaP layer grown from Sn-solution on a GaAs substrate; etching time 20 seconds

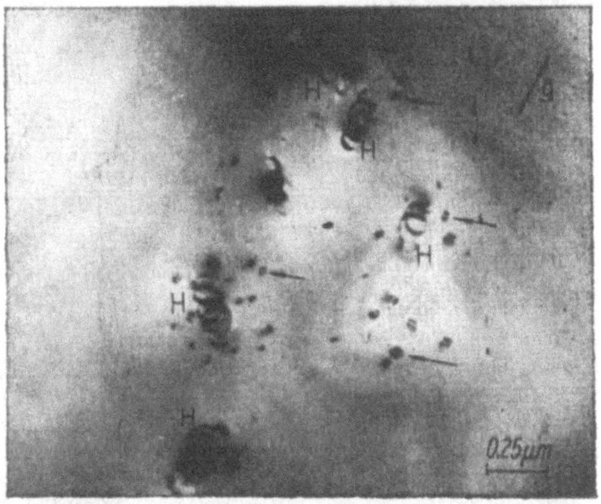

Fig. 5. TEM bright-field image (g = 220) of helices and microdefects observed in a GaP heteroepitaxial layer grown from Sn solution

positioned in the direct vicinity of the helices (Fig. 5 and 6b). The Figures 6a and b represent an exact one to one correlation for some defects. Similar to chapter 3.1. it is possible to compare the different items on the TEM images and optical micrograph.

3.3. LEC grown GaP simultaneously doped with S and Si

Figure 7 depicts a characteristic etch pattern obtained by the new etchant on LEC grown $(\bar{1}\bar{1}\bar{1})$ GaP. Besides well-defined D-pits (e.g. D in Fig. 7) a very high number of variously shaped small pits occurs always correlating to microdefects such as faulted and perfect dislocation loops and precipitates and/or inclusions. For example, small pits exhibiting a sharp apex can be seen. During continued etching those small pits shade into typical S-pits and vanish finally if the etching time is increased more and more. The Figures 8a–d represent the one to one correlation between etch pits and the true lattice defects. The bright halo around lattice defects is caused by the lower foil thickness resulting from etch pitting. In contrast to microdefects the inclined dislocations intersect both surfaces of the thin foil examined by TEM. Consequently two sharp etch pits respectively bright halos occur on the TEM micrograph (Figs 8c and d).

6a

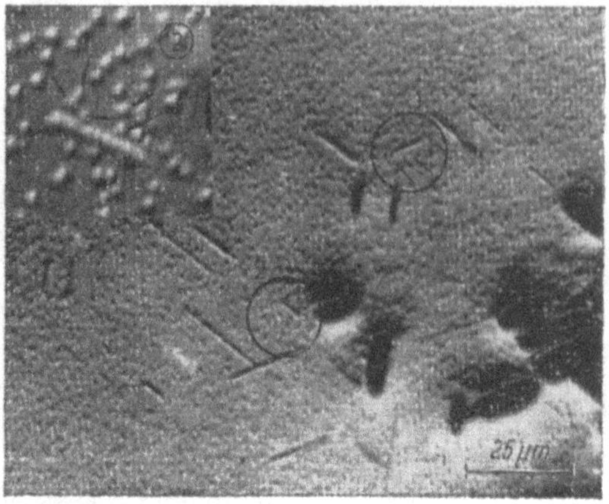

Fig. 6. a) Optical interference micrograph of etch features on a thinned an afterwards etched LPE grown GaP sample; b) TEM bright-field images of manyfold overlapping stacking faults and helices correlating to Figure 6a

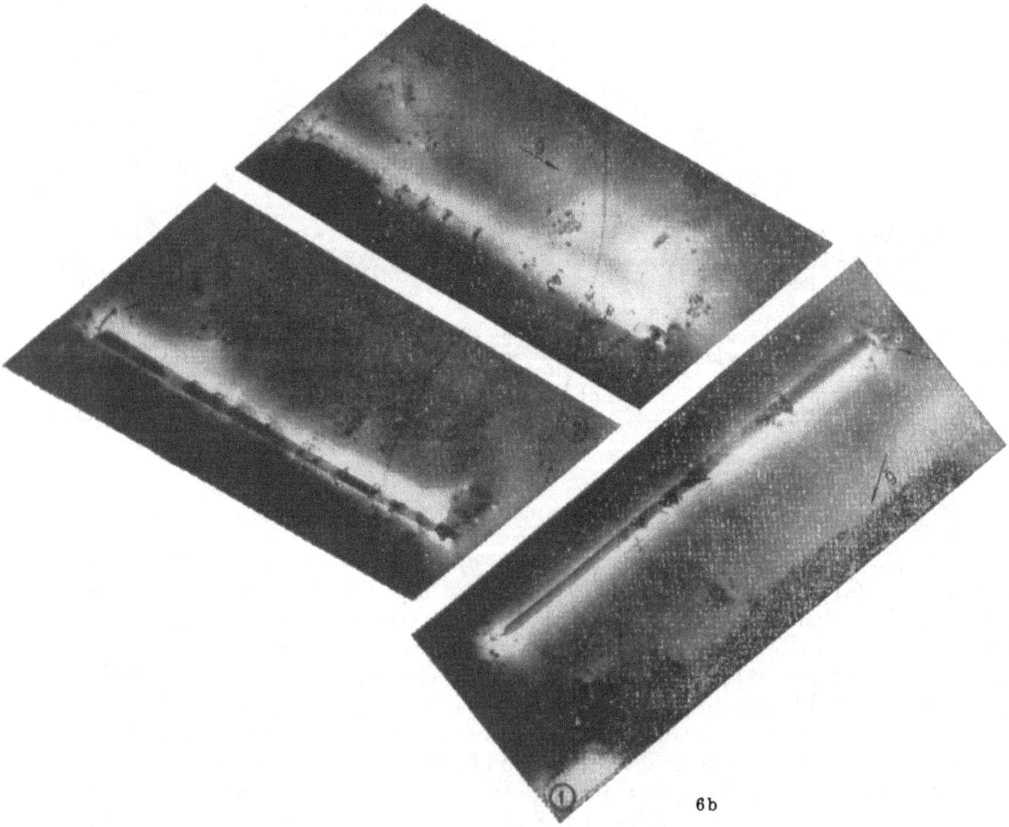

6b

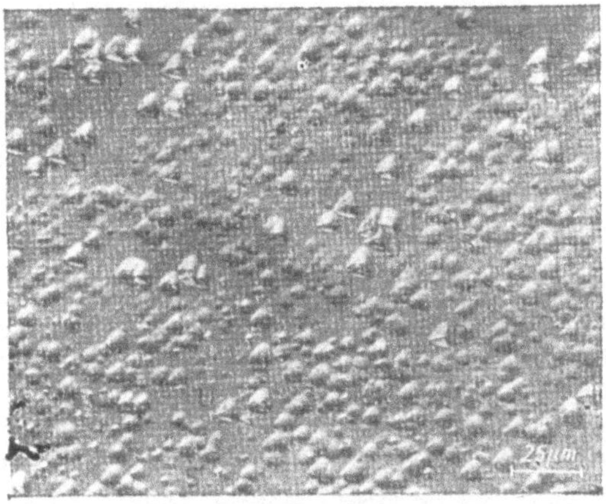

Fig. 7. Optical interference micrograph of etch pattern revealed on a (111) face of a LEC grown GaP sample simultaneously doped with sulphur and silicon; etching time 3 minutes; for explanation see text

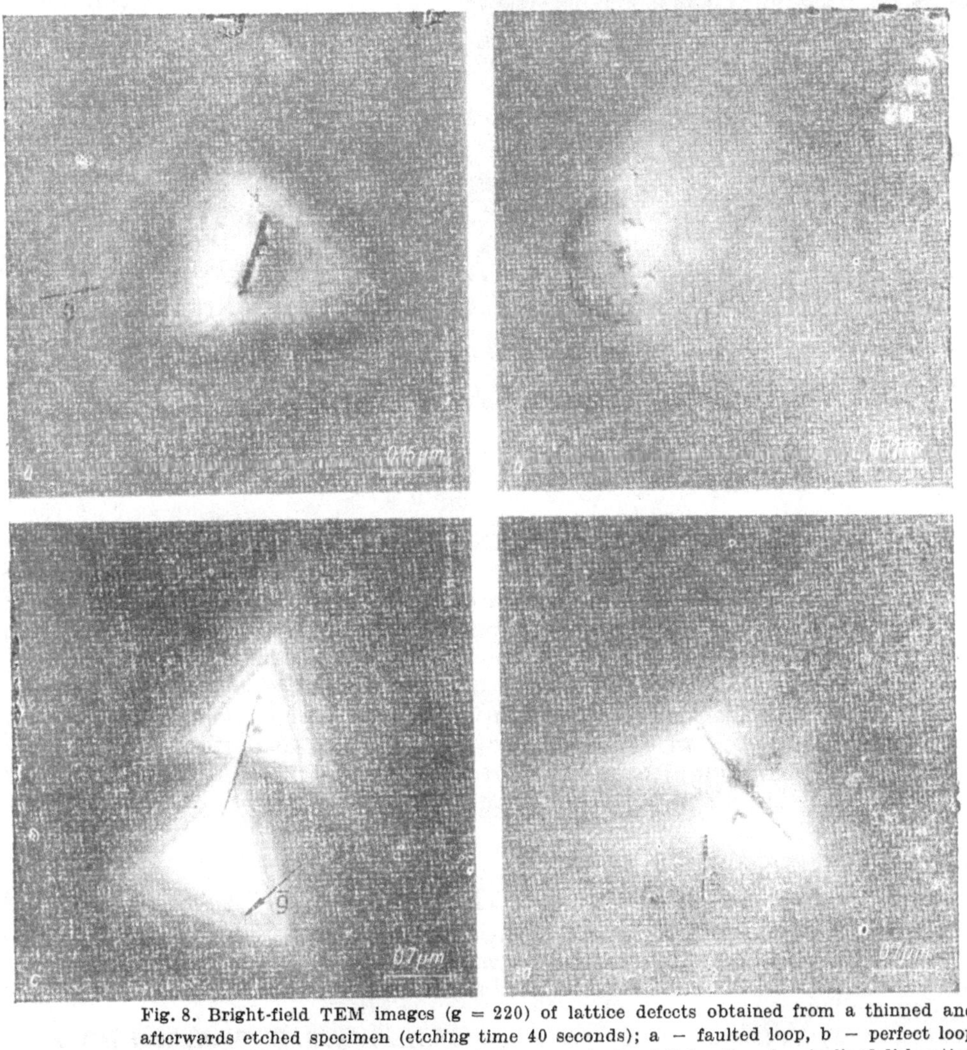

Fig. 8. Bright-field TEM images (g = 220) of lattice defects obtained from a thinned and afterwards etched specimen (etching time 40 seconds); a — faulted loop, b — perfect loop associated with precipitates and/or inclusions, c — inclined dislocation, d — inclined dislocation associated with microdefects

3.4. LEC grown InP doped with Sn

The new etchant produces well-defined D-pits (D and D_i in Fig. 9) on $(\bar{1}\bar{1}\bar{1})$ phosphorous faces of LEC grown InP too. Pits denoted with Di might be dislocations lying very flat with respect to the surface etched. During continued etching such D_i-pits do not disappear, but they change their locality appreciably. All of the other small pits occuring (examples are arrowed in Fig. 9) answer to very small defects as perfect dislocation loops, clusters of microdefects and so forth. During continued etching those small pits are transformed to typical S-pits (S in Fig. 9) which vanish finally. TEM investigations have manifested that this material contains besides grown-in dislocations large isolated perfect dislocation loops, clusters of faulted loops and precipitates and/or inclusions (WAGNER, GOTTSCHALCH 1986).

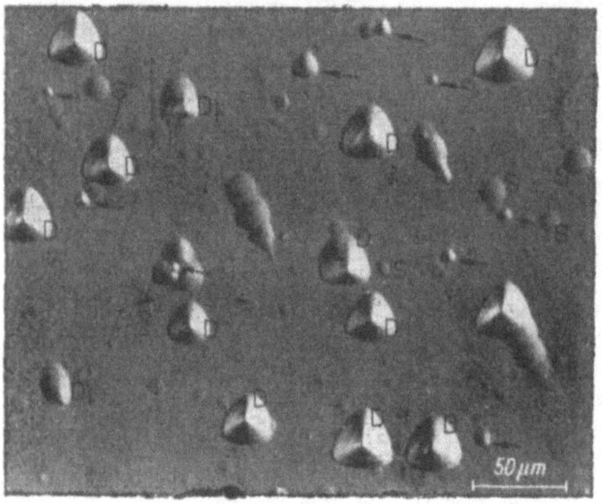

Fig. 9. Optical interference micrograph of an etched $(\bar{1}\bar{1}\bar{1})$ face of LEC grown InP doped with Sn; etching time 1 minute; for explanation see text

4. Summary

The etchant consisting of a mixture of HNO_3 and H_2O_2 produces well-defined etch pits and other explicable etch features on $(\bar{1}\bar{1}\bar{1})$ phosphorous faces of GaP and InP independent on the crystal growth procedure and doping. It is possible to distinguish between D-pits (grown-in dislocations), variously shaped small pits correlating to microdefects and grooves originated by stacking faults and microtwin lamellae. Subgrain twin boundaries could be also revealed but their etch features are more complicated. Compared to the other etchants mentioned in the introduction the new mixture discussed here exhibits an extremely low material removal during defect etching procedure. Therefore this etchant is favourable for characterization of very thin GaP and InP epitaxial layers of $(\bar{1}\bar{1}\bar{1})$ orientation.

References

ABRAHAMS, M., BUIOCCHI, C. J.: J. Appl. Physics **31**, 611 (1960)

AKITA, K., KOSUNOKI, T., KOMIYA, S., KOTANI, T.: J. Cryst. Growth **46**, 783 (1979)

BEPPU, T., TASHIRO, M., HIRAHARA, K., WATANABE, M.: Jpn. J. Appl. Physics **17**, 509 (1978)

BRANTLEY, W. A., LORIMOR, O. G., DAPKUS, P. D., HASZKO, S. E., SAUL, R. H.: J. Appl. Physics **46**, 2629 (1975)

BROWN, G. T., COCKAYNE, B., MCEWAN, W. R.: J. Mater. Sci. **15**, 2539 (1980)

Brown, G. T., Cockayne, B., McEwan, W. R.: J. Cryst. Growth **51**, 369 (1981)
Chu, S. N. G., Jodlauk, C. M., Ballman, A. A.: J. Electrochem. Soc. **128**, C 239 (1981)
Clarke, R. C., Robertson, D. S., Vere, A. W.: J. Mater. Sci. **8**, 1349 (1973)
Cockayne, B., Brown, G. T., McEwan, W. R., (a): J. Cryst. Growth **54**, 9 (1981)
Cockayne, B., Brown, G. T., McEwan, W. R., (b): J. Cryst. Growth **51**, 461 (1981)
Gottschalch, V., Kramer, P., Butter, E.: Cryst. Res. Technol. **13**, 543 (1978)
Huber, A., Linh, N. Z.: J. Cryst. Growth **29**, 80 (1975)
Iizuka, T.: J. Electrochem. Soc. **118**, 1190 (1971)
Komiya, S., Nakajima, K.: J. Cryst. Growth **48**, 403 (1980)
Kotani, T., Komiya, S., Nakai, S., Yamaoka, Y.: J. Electrochem. Soc. **127**, 2273 (1980)
Mullin, J. B., Royle, A., Stauchan, B. W.: Proc. Int. Symp. on Gallium Arsenide, Aachen 1970, (Inst. of Phys. Soc., London), p.
Neumann, H., Butter, E., Kramer, P., Ernst, H.-G., Kohl, W.: phys. stat. sol. (a) **55**, 99 (1979)
Richards, J. L., Crocker, A. J.: J. Appl. Physics **31**, 611 (1960)
Rosin, H., Freydank, G.: phys. stat. sol. (a) **32**, 32 (1975)
Rozgonyi, G. A., von Neida, A. R., Iizuka, T., Haszko, S. E.: J. Appl. Physics **43**, 3141 (1972)
Rozgonyi, G. A., Iizuka, T.: J. Electrochem. Soc. **120**, 673 (1973)
Siethoff, H.: phys. stat. sol. (a) **9**, 227 (1972)
Stringfellow, G. B., Lindquist, P. F., Cass, T. R., Burmeister, R. A.: J. Electron. Mater. **3**, 497 (1974)
Tajima, M.: Jpn. J. Appl. Phys. **16**, 1453 (1977)
Wagner, G., Kuehn, G., Pasemann, M.: Cryst. Res. Technol. **13**, 531 (1978)
Wagner, G., Kuehn, G.: J. Mater. Sci. Lett. **1**, 45 (1982)
Wagner, G., Gottschalch, V., (a): Pat. WP H01 L/247771 (1983)
Wagner, G., Gottschalch, (b): Cryst. Res. Technol. **21**, 881 (1980)
Werkhoven, C., Hengst, J. H. T., Bartels, W. J.: J. Cryst. Growth **42**, 632 (1977)
Werkhoven, C.: J. Electrochem. Soc. **125**, 671 (1978)
Werkhoven, C., van Odorp, C., Vink, A. T.: Philips Techn. Rev. **38**, 33 (1979)

(Received July 15, 1987)

Authors' address:

Dr. G. Wagner, Dr. V. Gottschalch
Karl-Marx-Universität, Sektion Chemie
Talstraße 35
Leipzig
DDR-7010

Cryst. Res. Technol.	23	1988	1	69–75

A. ZAHN, P. SCHREITER

Sektion Chemie der Karl-Marx-Universität, Wissenschaftsbereich Kristallographie
Leipzig, DDR

Lattice Constants and Site Preference in the System Ni_2SiO_4–Co_2SiO_4

Dedicated to Professor Hermann NEELS on the occasion of his 75th birthday

Most important supposition for a possible site preference in a olivine solid-solutions are besides the different crystal chemical properties of M cations (size, electronegativity, crystal field stabilization energy) the differences between the both octahedral coordinated sites M1 and M2 of the olivine structure. Using this regularities the site preference in the system Ni–Co olivines is predicted. If site preferences are existing, it is possible to deduce deviations of the Vegard's rule in the system Ni_2SiO_4–Co_2SiO_4 using geometrically-structurally considerations. The plot of experimentally determined lattice parameters in this solid solutions system shows the existing of the proposed deviations of the Vegard's rule and evidences the site preferences of Ni^{2+} in the M1- and Co^{2+} in the M2 octahedra only by determination of lattice parameters.

Die wesentlichsten Voraussetzungen für eine mögliche Lagenbevorzugung in Olivinmischkristallsystemen sind neben den kristallchemischen Eigenschaften der M-Kationen (Größe, Elektronegativität, Kristallfeldstabilisierungsenergie) die Unterschiede zwischen den beiden oktaedrisch koordinierten Lagen M1 und M2 der Olivinstruktur. In Anwendung dieser Gesetzmäßigkeiten wurde die Lagenbevorzugung im System Ni-Co-Olivin vorausgesagt. Bei Vorhandensein von Lagenbevorzugungen ist es in Olivinmischkristallsystemen und auch im System Ni_2SiO_4–Co_2SiO_4 möglich, Abweichungen von der Vegard-Regel nur aus geometrisch-strukturellen Überlegungen abzuleiten. Die graphische Darstellung der experimentell ermittelten Gitterkonstanten in diesem Mischkristallsystem zeigt das Vorhandensein der vorhergesagten Abweichungen von der Vegard-Regel und weist die Lagenbevorzugung von Ni^{2+} im M1- und Co^{2+} im M2-Oktaeder nur durch Bestimmung der Gitterkonstanten nach.

1. Introduction

Silicate olivines (variously symbolized A_2SiO_4 or M_2SiO_4) crystallize in the orthorhombic system, space group Pbnm. The most structural feature of the olivine structure represent the two distinct octahedral sites (M1 and M2), which are occupied by A cations. In this connection the aspect of site preference in olivine solid-solutions must be noted.

Many cation distribution studies of olivine solid solutions have been published during the last 15 years (ANNERSTEN et al.; BISH; BROWN, PREWITT; BROWN; FRANCIS, RIBBE; FUJINO; GAITE, HAFNER; GHOSE, WEIDNER; GHOSE, WAN; GHOSE et al.; LAGER, MEAGHER; NEWNHAM et al.; NORD et al.; ONKEN; OSTERTAG et al.; RAJAMANI et al.; SMYTH, HAZEN; WARNER, LUTH; WENK, RAYMOND; WILL, NOVER). The mostly used experimental methods are structure refinements by X-ray diffraction,

Mössbauer spectroscopy and spectra in the IR and visible region. The newer methods like ALCHEMI — Atom Location by Channeling, Enhanced Microanalysis (SPENCE, TAFTØ) and CHEXE — Channeling, Enhanced X-ray Emission are also applied successfully to the investigation of silicate olivines (SMYTH; TAFTØ, SPENCE). Neither lattice parameters nor experimental investigations of site preference are to be found for the binary olivine solid-solutions Ni_2SiO_4–Co_2SiO_4. We started therefore the synthesis of solid solutions and their endcomponents to investigate the lattice parameters for predicting the site preference in these solid solutions by the indirect way using only the lattice parameters.

2. Experimental and results

The synthesis of Ni_2SiO_4, Co_2SiO_4 and the solid solutions were made by solid state reactions:

- raw material: Co_3O_4, NiO, SiO_2 (amorphous)
- temperature of synthesis: 1200 °C
- runs of synthesis: Ni-olivine — 6 × 12 h sintering, ever 2 h activation; Co-olivine — 4 × 12 h sintering, ever 2 h activation; solid solutions like Ni-olivine.

The raw oxides were homogenized and pressed to pellets, which were placed on magnesia targets in the furnace. Between the runs the pellets were ground in a vibrator to homogenize and activate the material. After sintering the pellets were shocked in air. The completeness of reaction was proved by X-ray diffraction and microscopial techniques.

Diffuse reflection spectra in the range of 300 to 730 nm and IR-spectra (4000 — — 200 cm^{-1}) showed that cobalt and nickel occupy the octahedral sites of the olivine structure in the formal charge +2. The determination of solid solution components analogues to Mg–Mn–Fe olivines (BURNS, HUGGINS) and the forsterite-fayalite series (LEHMANN et al.) by these methods is not possible for this system.

The d-spacings we've got from X-ray powder patterns (Horizontalzählrohrgoniometer HZG 4, VEB Freiberger Präzisionsmechanik):

- $Cu_{K\alpha}$-radiation (quartz monochromator)
- internal standard — polycrystalline silicon
- step-by-step $2\theta = 0.005°$

Table 1
d-spacings (pm) in the olivine solid-solutions Ni_2SiO_4—Co_2SiO_4

hkl	olivine					
	$Ni_{2.0}$	$Ni_{1.6}$	$Ni_{1.2}$	$Ni_{0.8}$	$Ni_{0.4}$	$Co_{2.0}$
111	347.05	347.91	348.87	349.79	350.85	351.67
130	274.71	275.64	276.37	277.37	277.82	279.05
131	249.14	249.89	250.81	251.60	252.36	253.03
112	243.47	244.17	244.91	245.57	246.14	246.86
200	236.45	236.89	237.32	237.89	237.96	239.15
222	173.51	173.96	174.42	174.86	175.32	175.80
062	146.55	147.14	147.72	148.17	148.64	149.11

$Ni_{2.0}$–Ni_2SiO_4
$Ni_{1.6}$–$Ni_{1.6}Co_{0.4}SiO_4$
$Ni_{1.2}$–$Ni_{1.2}Co_{0.8}SiO_4$
$Ni_{0.8}$–$Ni_{0.8}Co_{1.2}SiO_4$
$Ni_{0.4}$–$Ni_{0.4}Co_{1.6}SiO_4$
$Co_{2.0}$–Co_2SiO_4

Table 2
Lattice parameters in the solid-solutions Ni_2SiO_4–Co_2SiO_4

olivine	a_0	b_0	c_0
Ni_2SiO_4	472.84(4)*)	1012.14(6)	591.64(5)
$Ni_{1.6}Co_{0.4}SiO_4$	473.74(4)	1016.50(6)	593.58(5)
$Ni_{1.2}Co_{0.8}SiO_4$	474.69(17)	1020.68(25)	595.87(22)
$Ni_{0.8}Co_{1.2}SiO_4$	475.64(5)	1023.88(8)	597.12(7)
$Ni_{0.4}Co_{1.6}SiO_4$	476.89(5)	1027.02(7)	598.95(6)
Co_2SiO_4	478.22(8)	1030.74(12)	600.30(11)

*) – Numbers in the parantheses are the calculated standard errors and refer to the last decimales

Table 1 shows the corrected d-spacings. The lattice parameters (Tab. 2, Fig. 3) were calculated by the least-squares refinement technique.

3. Discussions

3.1. Prediction of the intracrystalline cation distribution

Most important supposition for a possible site preference are besides the different crystal chemical properties of A-cations (in this case Co^{2+} and Ni^{2+}) the differences between the both octahedrally coordinated sites M1 and M2:

1. The M1 octahedron is smaller and more distorted than the M2 octahedron (Brown).
2. The M1 octahedron shares six and the M2 octahedron only three edges with neighbouring cation-occupied O-polyhedra (SiO_4 tetrahedra, M octahedra) (Brown).
3. The A-cation is able to realise a more covalent bonding in M1 than in the more ionic M2 site (Brown).

Besides the cation properties radius (Ni^{2+} – 69 pm; Co^{2+} – 74.5 pm (Shannon)) and electronegativity (Pauling) at the transition metal cations is need to take into considerations the crystal field stabilization energy (CFSE). Generally the CFSE depends on the crystal field splitting Δ and the stabilization during distortion σ (Burns). The crystal field splitting depends on the distances between cation M and ligand O ($\Delta \sim \sim 1/\langle M{-}O\rangle^5$). It is very important to note that σ is typically one to three orders of magnitude smaller than Δ (Bish).

Because of the electron configuration the CFSE of Ni^{2+} depends only on the size ratio of the both octahedra and the CFSE of Co^{2+} on the size ratio and distortion of the octahedra. Walsh et al. and Wood determined the site preference energies of M1 opposite to M2 in olivine to 1,9 kcal and 1,6 kcal, respectively (Ni^{2+}) and to 1.0 kcal (Co^{2+}) from spectroscopically measured crystal field splittings.

Using these facts may predict the following site preferences of Co^{2+} and Ni^{2+} in the discussed solid solutions:

1. The smaller cation Ni^{2+} has to prefer M1 and Co^{2+} M2. The electronegativities do not influence the site preference.
2. Ni^{2+} gains a larger CFSE in M1 relative to Co^{2+}. That means also Ni^{2+} has to prefer M1 and Co^{2+} M2.

3.2. Site preference and variation of lattice parameters

The lattice parameters perform the Vegard's rule if the cations Ni and Co are distributed statistically. The site preferences effect a deviation from this rule (Lumpkin, Ribbe).

It is possible to deduce such deviations in the solid solutions system Co–Ni olivine using geometrically-structurally considerations?

In the olivine structure exist chains of oxygen polyhedras, featured through edge-sharing between octahedra and tetrahedra. If only one type of octahedra take place in such a chain, then this type of octahedra determines the identity period of the chain. If such chains are running parallel crystallographic axis, the corresponding lattice parameters mainly depend on the cation occupation of these octahedras. This will be demonstrated with two projections of structure.

The projection down the a-axis shows zig-zag chains parallel [001], consisting of M1 octahedra and SiO_4 tetrahedra (Fig. 1a). They build layers parallel (001). M2 octahedra connect these layers. In this way edge-sharing chains of SiO_4 tetrahedra and M2 octahedra parallel to [010] are formed (Fig. 1b). Because of the existence of this chains we are able to predict a greater influence of M2 variation on the lattice parameter b_0 than M1 variation.

Figure 2a shows the projection of SiO_4 tetrahedra down the c-axis. To this projection M1- and M2 octahedra were added (Fig. 2b). We remark that parallel [100] the SiO_4 tetrahedra are edge-shared to chains only with M1 octahedra. Therefore it is

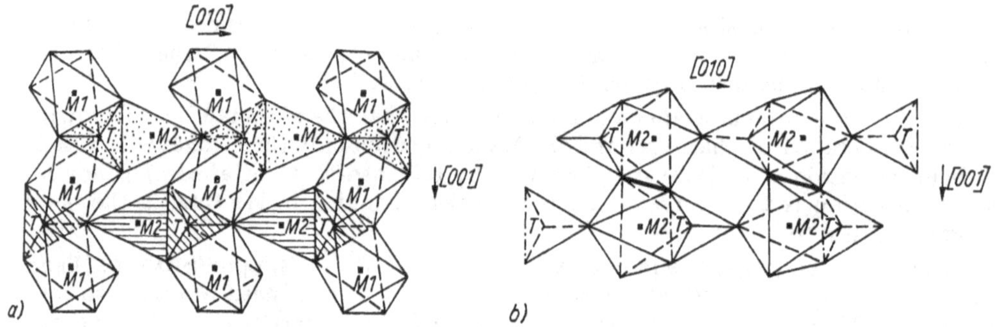

Fig. 1. a) Projection down the a-axis showing the M1 octahedra (■ M1) and the SiO_4 tetrahedra (T) of the olivine structure (—, – – planes of M2 octahedra and SiO_4 tetrahedra parallel to (100) and on a level); b) Projection down the a-axis showing chains of M2 octahedra and SiO_4-tetrahedra (– · – · – – edge-sharing between M2-octahedra and SiO_4 tetrahedra)

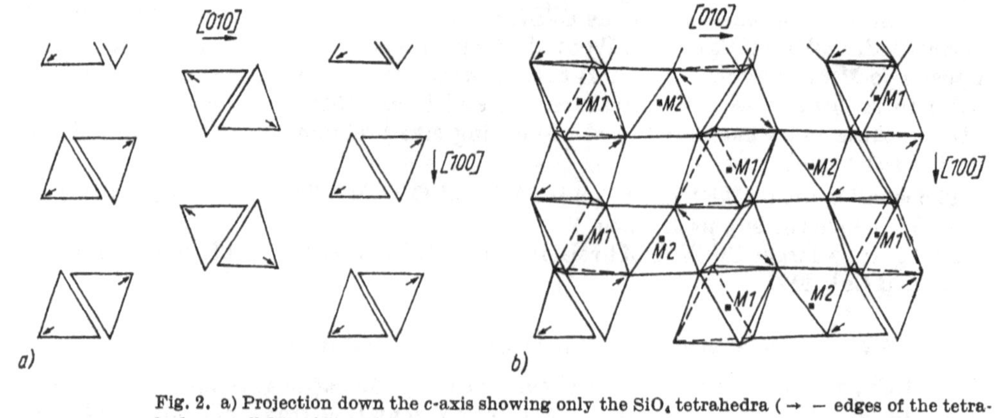

Fig. 2. a) Projection down the c-axis showing only the SiO_4 tetrahedra (→ – edges of the tetrahedra parallel to [001]; b) Projection down the c-axis showing the M1, M2 octahedra and the SiO_4 tetrahedra of the olivine structure

Table 3
Comparison of lattice parameters

olivine	a_0	b^0	$\Delta a/a \cdot 100\%$	$\Delta b/b \cdot 100\%$
Mg_2SiO_4 (Smyth, Hazen)	4.756	10.207		
$CaMgSiO_4$ (Onken)	4.822	11.108	1.4	8.5
γ-Ca_2SiO_4 (Brown 1982)	5.078	11.225	5.2	1.1

possible to predict, that the lattice parameter a_0 more depends on variation of M1- than M2 octahedra.

To find an answer of the question "Whether the variation of kinds of octahedra results in the same quantity of the variation of lattice parameters a_0 and b_0?", we use an example.

We consider variation of lattice parameters occupying M1- and M2 octahedra by the same cations (Tab. 3) and the kind of sharing in the octahedra-tetrahedra chains. The percentage differences of lattice parameters ($\Delta a/a \cdot 100\%$; $\Delta b/b \cdot 100\%$) of forsterite (Mg_2SiO_4) and monticellite ($CaMgSiO_4$), respectively (only the M2 occupation varied) are in b_0 8.5% and in a_0 1.4% only. Relative to monticellite and γ-Ca_2SiO_4 (variation only in M1) these values amount to a_0 5.2% and to b_0 1.1%. That means the variation only in M2 ($Mg_2SiO_4 \rightarrow CaMgSiO_4$) causes a difference between $\Delta b/b \cdot 100\%$ and $\Delta a/a \cdot 100\%$ of 7.1% whereas the analogous difference in respect of M1 occupation ($CaMgSiO_4 \rightarrow \gamma$-$Ca_2SiO_4$) is only 4.1%. Summarized, this shows, that if variated the occupation of different octahedra the change of M2 octahedra causes a larger difference in b_0 than M1 in a_0, and also in the deviation of the Vegard's rule. The structural reasons lies in the octahedra-tetrahedra linkage in the chains (see Figs 1 and 2), where in the M1 octahedra two edges are fixed by the inflexible SiO_4 tetrahedra and in the M2 octahedra only one, resulting in different variations of octahedra volumes M1 and M2, respectively, if the same cation exchange takes place.

4. Summary

The deviations of the Vegard's rule are derivable from as follows:

1. If ordering cation distribution (smaller cation prefers M1) lattice parameter b_0 is larger and a_0 is smaller than without site preference.
2. If anti-ordering cation distribution (smaller cation prefers the bigger M2) lattice parameter b_0 becomes smaller and a_0 becomes larger than statistical cation distribution.

The plot of lattice parameters in the solid-solutions Ni_2SiO_4–Co_2SiO_4 (Fig. 3) shows:

1. A deviation of the Vegard's rule is existing.
2. The deviation from the Vegard's rule in a_0 and b_0 is contrary. That means Co and Ni in this solid solutions show site preference.
3. In the solid-solutions Ni_2SiO_4–Co_2SiO_4 Ni^{2+} prefers the M1-octahedra and Co^{2+} prefers the M2-octahedra, in application of the previous structural discussions.

With this paper we show the possibility to prove experimentally the site preference of cations in olivine solid solutions only by determination of lattice parameters.

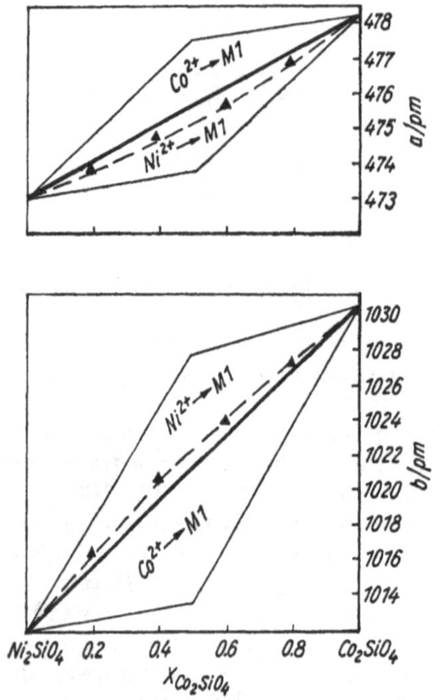

Fig. 3. A plot of the lattice parameters in the olivine solid-solutions (Ni_2SiO_4–Co_2SiO_4) (▲ – experimentally determined lattice parameters; △ – constructed areas of deviations from the Vegard's rule using the equations of LUMPKIN and RIBBE if M1-site preferences of Co andNi are existing)

References

ANNERSTEN, H., ERICSSON, T., FILLIPIDIS, A.: Amer. Mineral. **67**, 1212 (1982)
BISH, D. L.: Amer. Mineral. **66**, 770 (1981)
BROWN, G. E., PREWITT, C. T.: Amer. Mineral. **58**, 577 (1973)
BROWN, G. E.: Olivines and Silicate Spinels in Reviews in Mineralogy (vol. 5), Orthosilicates, Second Edition, Mineralogical Soc. Amer., P. H. RIBBE (Ed.) 1982
BURNS, R. G.: Amer. Mineral. **55**, 1608 (1970)
BURNS, R. G., HUGGINS, F. E.: Amer. Mineral. **57**, 967 (1972)
FRANCIS, C. A., RIBBE, P. H.: Amer. Mineral. **65**, 1263 (1980)
FUJINO, K.: Kobutsugaku Zasshi (Jap.) **14**, 135 (1980)
GAITE, J. M., HAFNER, S. S.: J. Chem. Phys. **80**, 2747 (1984)
GHOSE, S., WAN, C.: Contrib. Mineral. Petrol. **47**, 131 (1974)
GHOSE, S., WEIDNER, J. R.: Geol. Soc. Amer., Annual Meetings, Miami Beach, Florida, Abstracts with Programs, **6**, 751 (1974)
GHOSE, S., WAN, C., OKUMURA, F. P., OHASHI, H., WEIDNER, J. R.: Acta Cryst. A **31**, Suppl. S 76 (1976)
LAGER, G. A., MEAGHER, E. P.: Amer. Mineral. **63**, 365 (1978)
LUMPKIN, G. R., RIBBE, P. H.: Amer. Mineral. **68**, 164 (1983)
NEWNHAM, R., CARON, L. G., SANTORO, R.: J. Amer. Ceram. Soc. **49**, 284 (1966)
NORD, A. G., ANNERSTEN, A., FILIPPIDIS, A.: Amer. Mineral. **67**, 1206 (1982)
ONKEN, H.: Tschermak's Mineral. Petrogr. Mitteilungen **10**, 34 (1965)
OSTERTAG, R., AMTHAUER, G., RAGER, H., McSWEEN, H. Y.: Earth. Planet. Sci. Letters **67**, 162 (1984)
PAULING, L.: Die Natur der chemischen Bindung, Weinheim 1964
RAJAMANI, V., BROWN, G. E., PREWITT, C. T.: Amer. Mineral. **60**, 292 (1975)
SHANNON, R. D.: Acta Cryst. A **32**, 751 (1976)
SMYTH, J. R., HAZEN, R. M.: Amer. Mineral. **58**, 588 (1973)
SMYTH, J. R.: Geophys. Res. Letters **9**, 1113 (1982)

SPENCE, J. C. H., TAFTØ, J.: J. Microscopy **130**, Part 2, 147 (1983)
TAFTØ, J., SPENCE, J. C. H.: Science **218**, 59 (1982)
WALSH, D., DONNAY, G., DONNAY, J. D. H.: Bull. Soc. Franc. Mineral. Crist. **97**, 170 (1974)
WARNER, R. D., LUTH, W. C.: Amer. Mineral. **58**, 998 (1973)
WENK, H. R., RAYMOND, K. N.: Z. Krist. **137**, 86 (1973)
WILL, G., NOVER, G.: Phys. Chem. Minerals **4**, 199 (1979)

(Received July 9, 1987)

Authors' address:

Doz. Dr. P. SCHREITER, Dr. A. ZAHN
Karl-Marx-Universität, Sektion Chemie, Wissenschaftsbereich Kristallographie
Scharnhorststr. 20
Leipzig
DDR-7030

Cryst. Res. Technol.	23	1988	1	77–84

M. Schenk, H. Berger, A. Klimakow, M. Mühlberg, M. Wienecke

Humboldt-Universität zu Berlin, Sektion Physik, Bereich Kristallographie, Berlin, DDR

Nonstoichiometry and Point Defects in PbTe

Dedicated to Professor Hermann Neels on the occasion of his 75th birthday

The nonstoichiometry and point-defect structure of PbTe was investigated by various methods including carrier-concentration, lattice-constant, positron-annihilation and vapour-density measurements. The results are discussed in comparison with data published by other authors. New aspects are obtained especially from high-temperature experiments. A model of nonstoichiometry and defect structure in PbTe is proposed. The stability region of PbTe is essentially broader than reported until now. The disorder takes place mainly in the Pb sublattice, i.e., there occur charged Pb vacancies (Te-rich) and Pb interstitials (Pb-rich), respectively, as well as probably additional neutral Pb interstitials in Pb-rich PbTe.

Die Nichtstöchiometrie und Punktdefektsstruktur von PbTe wurde mit verschiedenen Methoden untersucht, u. a. durch Messungen der Ladungsträgerkonzentration, der Gitterkonstante, der Positronenannihilation sowie der Dampfdichte. Die Ergebnisse werden mit denen anderer Autoren verglichen. Neue Aspekte ergeben sich insbesondere aus Hochtemperaturmessungen. Ein Modell der Nichtstöchiometrie und der Defektstruktur in PbTe wird vorgeschlagen. Das Stabilitätsgebiet von PbTe ist wesentlich breiter als bisher berichtet. Die Fehlordnung tritt hauptsächlich im Pb-Untergitter auf, d. h., es existieren geladene Pb Leerstellen (Te-reich) bzw. Pb-Zwischengitterionen (Pb-reich) sowie zusätzlich wahrscheinlich neutrale Pb-Zwischengitteratome in Pb-reichem PbTe.

1. Introduction

Binary compounds are not strongly stoichiometric at $T > 0$ K. Correlating with the band gap and bonding type, and depending on temperature, pressure, etc., they exhibit finite existence ranges, i.e., they exist as a single phase in certain composition intervals. This nonstoichiometry can be written as $A_{1-y}B_y$, with $y = 0.5 \pm \Delta y$. In general, insulators and semiconductors have narrow existence ranges and metallic compounds have wide ones (Albers, Haas). On the atomic scale, the deviation from stoichiometry is realized by point defects (vacancies, interstitials, and anti-site defects respectively). In semiconductors, the existence of point defects leads generally to free carriers. In this way, the physically and technically relevant properties of semiconductors are decisively determined by nonstoichiometry. In cases of extremely small deviations from stoichiometry, the measurement of the carrier concentration is mostly the only method to investigate this deviation.

PbTe, a semiconductor with a small band gap, crystallizing in the NaCl structure, is interesting as a basic material for infrared optoelectronic devices. The maximal width of the stability region, calculated from carrier concentration measurements, has been reported to be $\Delta y/y \approx 10^{-4}$ (Kovalchik, Maslakovets; Brebrick, All-

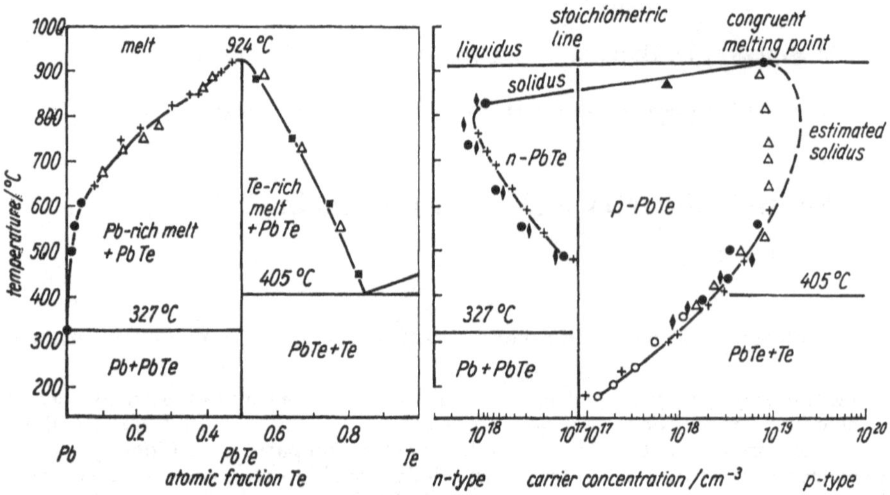

Fig. 1. Phase diagram of the system Pb–Te (left) and the stability region of PbTe (right). Data from MILLER and KOMAREK +, LUGSCHEIDER et al. △, HARRIS et al. ●, LAUGIER ■ (left side); SCANLON ○, BREBRICK and GUBNER △, HARMAN ●, STRAUSS ▽, HEWES et al. + (righ side)

GAIER; BREBRICK, GUBNER; MILLER, KOMAREK; SCANLON; HARMAN; SEALY, CROCKER; HEWES et al.; STRAUSS; GEORGE, WAGNER), assuming a charge of 1 (HARMAN; SEALY, CROCKER; GEORGE, WAGNER) or 2 per defect (PARADA, PRATT; LEUTE, SCHMIDTKE; FUJIMOTO, SATO; BREBRICK). Usually, the investigations of the stability region are performed at room temperature, the measuring samples quenched from high-temperature equilibrium state.

Figure 1 shows some compiled literature data of temperature versus composition measurements (Δy as carrier concentration). A Te surplus acts as acceptors ("p-type"), a Pb excess as donors ("n-type"). The temperature-independent course of the p-side curve at $p \approx 10^{19}$ cm^{-3} and $T_c \gtrsim 525$ °C is explained by the poor heat conductivity of the material, i.e., quenching from above a critical temperature T_c does not conserve the equilibrium state, but the excess material precipitates as a second phase, according to the retrograde course of the solidus line (ALBERS et al.). This precipitation could be confirmed at the first time by our TEM investigations (MÜHLBERG, HESSE).

In spite of the probable variation of the defect state during quenching from high temperature, due to precipitation, most experiments in recent years have been made on those samples, e.g. by carrier-concentration and lattice-constant measurements. The alternative are high-temperature in-situ investigations, e.g. of carrier concentrations (ALBERS, WIENECKE) and of optical vapour densities (BREBRICK, STRAUSS), which allow the calculation of nonstoichiometry and partial pressure.

Because of the importance of the defects due to non-stoichiometry in PbTe concerning the growth and properties of the crystals, and on account of the contradictory results and models concerning the stability region, we represent and discuss in the following some experimental results in order to contribute to the better understanding of these problems. Various methods, completing one another, which give informations about resulting strain fields (X-ray diffraction) as well as about charged point defects (carrier concentration, positron annihilation), have been employed.

2. Experimental

PbTe crystals of at least 99.999% purity were grown by the vertical Bridgman technique. About 1 mm thick slices of these crystals were annealed at different temperatures and partial pressures of the constituents for varying the native defect density within the stability region. After reaching equilibrium conditions the ampoules containing the samples were quenched in air or in water. The samples have been investigated by means of X-ray precision lattice-constant measurements using the Soller-slit method (BERGER), by Hall-coefficient and conductivity measurements, using the van-der-Pauw method as well as the hot-probe technique, and in special cases by positron-annihilation experiments (DLUBEK et al. 1987). Positron annihilation is a modern method for determination of vacancies acting as acceptors in semiconductors (DLUBEK et al. 1986). Besides of that, electrical in-situ measurements in dependence on temperature have been made. The equipment for annealing samples up to 800 °C and for Hall-effect and conductivity measurements according to van der Pauw is described by ALBERS and WIENECKE. Some vapour density investigations according to the method of BREBRICK and STRAUSS were performed at higher temperatures.

3. Results and discussion

3.1. Stability region

Several efforts were made to determine the width of the stability range of PbTe more exactly. Special quenching arrangements served for a relatively good maintenance of the high-temperature equilibrium state due to different annealing conditions. We assume especially that in the surface region of the samples the cooling velocity was sufficient to retain a great part of the single phase non-stoichiometrically, i.e., the point defects should be evident from surface investigation methods, e.g. X-ray and hot-probe measurements. In these cases the depth contributing to the measurement is in the order of some ten microns.

The spread of the experimental results was nevertheless great, probably due to inhomogeneities as a result of an uneven phase boundary during growth of the crystal from the melt. Another reason could be, of course, the hardly reproducible quenching conditions. Moreover, on the Pb-rich side the results changed as a function of time after quenching, possibly due to diffusion effects at room temperature (BERGER et al.). In spite of these difficulties, the measurements showed a broader existence range than reported until now: on the Te-rich side at least to $p = 2 \times 10^{19}$ cm^{-3}, on the Pb-rich side at least to $n = 8 \times 10^{18}$ cm^{-3}.

These results are supported by high-temperature in-situ measurements of the optical vapour density. At equilibrium conditions the partial pressures of the constituents in a closed annealing ampoule are directly proportional to the composition in the solid (BREBRICK, STRAUSS). After calibration of the light absorption by the vapour, we found the width of the Te-rich side of the existence region to be at least $p = 2 \times 10^{19}$ cm^{-3} at 800 °C.

More exact conclusions are possible after high-temperature in-situ measurements of Hall coefficient and conductivity (WIENECKE, SCHENK). Single crystalline Hall rods of $5 \times 5 \times 1$ mm^3 size were used for isochronic temperature-time experiments. The temperature levels were held during a defined time (24 h), in which the electrical properties were analyzed. The measurements have been made with a great number of samples and with different temperature steps. The results illustrated in Figure 2 are typical for all samples investigated. The arrows show the direction of temperature change. Such measurements are possible, because the conductivity due to the extrinsic carriers is much higher than the intrinsic conductivity.

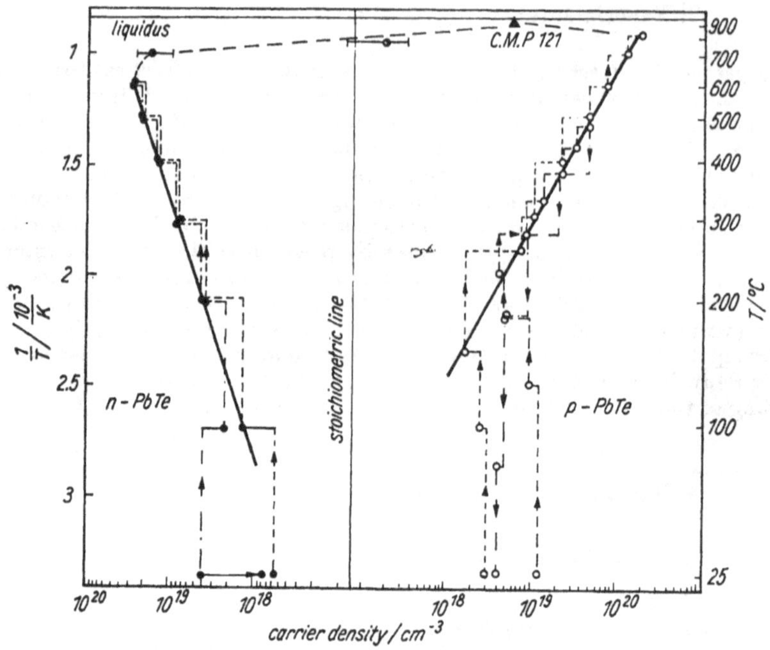

Fig. 2. Composition limits of stability of PbTe given by in-situ measurements. Explanation in the text

The stability region is substantially broader than reported in the literature up to now (cf. Fig. 1). It extends to a carrier concentration $p = 2 \times 10^{20}\ \mathrm{cm^{-3}}$ at 800 °C in Te-rich and to $n = 3 \times 10^{19}\ \mathrm{cm^{-3}}$ at 600 °C in Pb-rich PbTe, respectively. We are not sure that these values correspond to deviations from stoichiometry (concentrations of 10^{-2} and 3×10^{-3}, resp., assuming singly charged defects), because additional, electrically neutral defects cannot be excluded (see 3.2). Furthermore, from Figure 2 can be derived that at higher temperatures there exist several oppositely charged defects, which compensate one another electrically. This problem should be resolvable only by combined high-tempreature X-ray and dilametric experiments (Simmons, Balluffi). In spite of all difficulties, the relatively great nonstoichiometry should encourage to perform such investigations.

3.2. Kind of defects

Although it has been shown in 3.1 that experiments with quenched samples are not sufficient for understanding the high-temperature state of defect structure, we presume that not the kind of defects (vacancies, interstitials, anti-site defects), but only their number is changed by quenching. Therefore, we made precision lattice-constant measurements of differently heat treated PbTe samples. As a measure of the number of defects serves the carrier concentration, investigated by means of the hot probe technique. At the p-type side, the lattice constant tends to decrease with increasing carrier concentration (Fig. 3), while in Pb-rich PbTe an increase of the lattice constant with increasing carrier concentration could be ensured because of the small concentration change only by statistical analysis of the experimental data.

In general, vacancies should lead to decreasing lattice parameters, whilst interstitials should cause an increase of the lattice constant, at least for relatively close-packed structures. In compounds with ionic bonding portions like PbTe, vacancies could also cause a repulsion of the six octahedrally arranged nearest neighbours of them, because all have the same charge. That would lead to an increasing lattice constant, as has been pointed out by DRISCOLL and WILLOUGHBY in the case of GaAs, which is, however, tetrahedrally bound. The above hypothesis is certainly only valid, if no opposite charges (electrons, holes) are trapped in the vacancies, compensating the absent ion and avoiding the coexistence of equally charged ions. In the latter case the model of decreasing lattice constants should be valid. On the other hand, the decreasing lattice parameter at Te excess can only be interpreted assuming Pb vacancies V_{Pb}^- or $V_{Pb}^=$. An anti-site defect Te_{Pb}^- is improbable after our opinion, because the Te anti-site ion, surrounded by six negatively charged Te ions, would have to deliver one or two electrons from its outer shell, which cannot be supposed. So we conclude in agreement with many other authors (e.g., SEALY, CROCKER) that the predominating kind of defect on the Te-rich side of the existence region in PbTe is a singly or doubly ionized Pb vacancy.

To ensure this conclusion we made positron-annihilation measurements of p-type PbTe (DLUBEK et al. 1987), which confirmed the assumption of negatively charged Pb vacancies.

Corresponding to Figure 2, we presume that V_{Pb}^- is the predominant defect of p-type PbTe up to 800 °C. From the slope of the reciprocal temperature dependence on lg p, the formation energy E^f can be calculated to $E^f = (0.32 \pm 0.05)$ eV. This value is well comparable with that of KOBAYASHI et al. who determined 0.3 eV for the formation energy of a Pb vacancy. The motion energy E^m was evaluated to be $E^m = (0.62 \pm 0.05)$ eV, which agrees well with $E = 0.6$ eV, found by BOLTAKS for the activation energy of Pb diffusion via Pb vacancies in Te-rich PbTe.

As can be seen from Figure 2, also the Pb-rich side of the stability region of PbTe is determined by one kind of charged point defect up to 600 °C. Corresponding to the tendencies of lattice constant changes, Te vacancies V_{Te}^+ or V_{Te}^{++} can be very probably eliminated. Stable anti-site defects Pb_{Te}^+ or Pb_{Te}^{++} cannot be supposed for analoguous reasons as discussed previously for Te-rich PbTe. Conversely, Pb interstitials Pb_i acting as donors can explain all the experimental results, the expanded lattice as well as the positron-annihilation results (no defect signal, because interstitials and positively charged vacancies cannot act as traps for positrons (DLUBEK et al. 1987)). The formation and motion energies, calculated from Figure 2, are surprisingly small: $E^f = (0.09 \pm 0.05)$ eV, $E^m = (0.18 \pm 0.05)$ eV. They differ extremely from literature data ($E^f = 1.2$ eV (KOBAYASHI et al.), $E^m = 1.04$ eV (GOMEZ et al.)), but they explain the previously mentioned room-temperature time-dependent changes of lattice constants and carrier concentrations. The latter fact is also visible from Figure 2: the carrier concentration of the n-PbTe samples, cooled down to room temperature, was reduced from 4×10^{18} cm^{-3} to 8×10^{17} cm^{-3} after one day. Only a defect with such a very low motion energy is able to diffuse at room temperature with measurable effect.

These considerations agree with the basic statement of BREBRICK and STRAUSS that the Te sublattice of PbTe is such more stable than the Pb sublattice. So, one can conclude that the main charged defects exist in the Pb sublattice: on the Pb-rich side of the stability region predominate Pb interstitial ions Pb_i, on the Te-rich side Pb vacancies V_{Pb}. These considerations are valid relating to the carrier-concentration measurements. The effect of charged foreign atoms on the electrical investigations can be neglected because of the high purity of the material. But, in addition, one

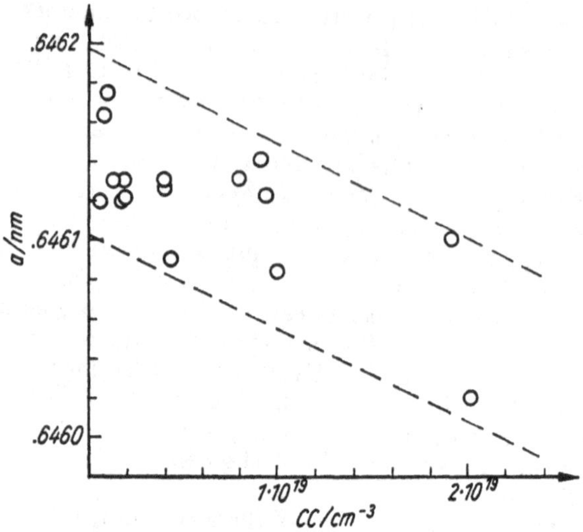

Fig. 3. Lattice constants a versus carrier concentration (cc) of p-PbTe

Table 1
Comparison of experimental and calculated lattice-constant changes for n-type PbTe. I for purely ionic bonding, II for mixed covalent-ionic bonding (40% ionic)

No. (pair of measuring values)	change of defect concentration [10^{-17} cm^{-3}] exp.	change of lattice constant [10^7 nm]		
		exp.	calc.	
			I	II
1	(4.5 ± 1.5)	110 ± 40	2.2 ± 0.4	2.8 ± 0.6
2	9	80 ± 40	3.9	5.1
3	5	60 ± 50	2.2	2.8

has to take into account neutral defects, not influencing the carrier concentration. Therefore, we have tried to evaluate quantitatively the relationship between lattice-constant changes and point-defect concentrations on the basis of a simple continuum model of the defect-induced lattice relaxation (see, e.g., PAUFLER, SCHULZE). For the calculation we assumed that the defect concentration corresponds to the measured carrier concentration. Whilst the calculated lattice-constant changes for p-type PbTe are in the same order of magnitude as the observed ones (cf. Fig. 3), the values for n-type specimens show a striking discrepancy. Some examples of lattice-constant changes measured on n-type PbTe in comparison with calculated values are given in Table 1. Independently of the radii used for calculation, the experimentally found changes are more than one order of magnitude greater than the calculated values. This can be explained by an essentially higher total point-defect concentration than those detected by carrier measurements. Hence, besides of charged Pb interstitials, at least one additional kind of point defect must exist. Because other kinds of charged defects occur only at high temperatures in a sufficient number, as results from the in-situ investigations (cf. 3.1), the existence of additional neutral Pb interstitials is supposed. If this conclusions is true, the Pb-rich side of the stability region should be still more extended than derived from the electrical measurements.

3.3. Precipitation of excess defects

According to the retrograde slope of the solidus line, outside of the stability range at lower temperatures the defects coagulate and precipitate. As we could show by TEM measurements of Te-rich PbTe (Mühlberg, Hesse), Te coagulates as very small needles with axes parallel to ⟨100⟩. In our opinion, this precipitation takes place by Pb outdiffusion from certain areas via a Pb vacancy mechanism, producing Te-rich zones, i.e. the "precipitates". This process could also be expressed as a Pb vacancy agglomeration.

For Pb-rich PbTe no adequate TEM measurements have been done. We only assume that at room temperature in n-type crystals a high density of very small Pb precipitates exists as a consequence of low motion energy of Pb interstitials. The Pb diffusion on the n-type side is of interstitial kind, therefore, the Pb-rich precipitates have to be high-density conglomerates of Pb and Te at PbTe lattice sites and, additionally, Pb at interstitial sites. Nothing can be said about the contribution of such second phase precipitates to the lattice constant of the basic phase; but they should, however, contribute to the carrier concentration.

4. Conclusions

From our investigations we propose the following model of nonstoichiometry and point-defect structure of PbTe:

— The stability region of PbTe is more than one order of magnitude larger than known from literature.

— One charged defect kind is predominant up to 800 °C in p-type PbTe and another kind up to 600 °C in n-type material. At higher temperatures, there must coexist various kinds of defects, compensating each other electrically.

— The disorder takes place mainly in the Pb sublattice, i.e., the predominating point defects are V_{Pb} and Pb_i, respectively. Probably in p-type PbTe there exist predominantly charged Pb vacancies, in n-type material charged and neutral interstitials.

— Precipitation occurs by means of the Pb vacancy mechanism in p-type PbTe, and by interstitial diffusion of Pb_i in n-PbTe.

We are indebted to Professor G. Dlubek and Dr. R. Krause for positron-annihilation investigations, to Dr. C. Albers and Dr. R. Behrendt for electrical measurements, and to Mr. U. Menzel and Mr. A Lehmann for X-ray experiments.

References

Albers, C., Wienecke, M.: Cryst. Res. Technol. **22**, K53 (1987)
Albers, W., Haas, C.: Philips Techn. Rdsch. **30**, 74 (1969/70)
Albers, W., Haas, C., Vink, H. J.: Philips Res. Rep. **18**, 372 (1963)
Berger, H.: J. Appl. Cryst. **17**, 451 (1984)
Berger, H., Lehmann, A., Schenk, M.: Cryst. Res. Technol. **20**, 579 (1985)
Boltaks, B. I.: Zh. Tekh. Fiz. **26**, 2448 (1956)
Brebrick, R. F.: J. Electr. Mater. **6**, 659 (1977)
Brebrick, R. F., Allgaier, R. S.: J. Chem. Phys. **32**, 1826 (1960)
Brebrick, R. F., Gubner, E.: ibid. **36**, 1283 (1962)
Brebrick, R. F., Strauss, A. J.: ibid. **40**, 3230 (1964)
Dlubek, G., Brümmer, O., Plazaola, F., Hautojärvi, P.: J. Phys. **C19**, 331 (1986)

Dlubek, G., Krause, R., Mühlberg, M., Schenk, M.: phys. stat. sol. (a) (in preparation, 1987)
Driscoll, C. M. H., Willoughby, A. F. W.: Proc. Int. Conf. "Rad. Damage and Defects in Semicond.", Conf. Ser. Nr. 16, Reading (1972), The Inst. of Physics, London, p. 377
Fujimoto, M., Sato, Y.: Jap. J. Appl. Physics **5**, 2 (1966)
George, T. D., Wagner, J. B.: J. Phys. Chem. Solids **30**, 2359 (1969)
Gomez, M. P., Stevenson, D. A., Huggins, R. H.: ibid. **32**, 335 (1971)
Harman, T. C.: J. Nonmetals **1**, 183 (1973)
Harris, J. S., Longo, J. T., Gertner, E. R., Clarke, J. E.: J. Cryst. Growth **28**, 334 (1975)
Hewes, C. R., Adler, M. S., Senturia, S. D.: J. Appl. Physics **44**, 1327 (1973)
Kobayashi, I., Sato, A., Fujimoto, M.: Proc. 7. Int. Conf. Physics of Semicond., Paris 1964, Ed. M. Hulin, 1964, p. 1257
Kovalchik, T. I., Maslakovets, Yu. P.: Sov. Phys. Tech. Phys. **1**, 2337 (1956)
Laugier, A.: Rev. Phys. Appl. **8**, 259 (1973)
Leute, V., Schmidtke, H.: Ber. Bunsenges. **79**, 1135 (1975)
Lugscheider, W., Ebel, H., Langer, G.: Z. Metallkde. **56**, 581 (1965)
Miller, E., Komarek, K.: Trans. MS AIME **236**, 832 (1966)
Mühlberg, M., Hesse, D.: phys. stat. sol. (a) **76**, 513 (1983)
Parada, N. J., Pratt, G. W.: Phys. Rev. Lett. **22**, 5 (1969)
Paufler, P., Schulze, G. E. R.: Phys. Grundlagen mechan. Festkörpereig., Bd. 1, Berlin 1978, p. 65
Scanlon, W. W.: Phys. Rev. **126**, 509 (1962)
Sealy, B. J., Crocker, A. J.: J. Mater. Sci. **8**, 1731 and 1737 (1973)
Simmons, R. O., Balluffi, R. W.: Phys. Rev. **117**, 52 (1960)
Strauss, A. J.: J. Electr. Mater. **2**, 553 (1973)
Wienecke, M., Schenk, M.: phys. stat. sol. (a) **101**, 676 (1987)

(Received July 15, 1987)

Authors' addresses:

Prof. Dr. habil. Manfred Schenk, Doz. Dr. sc. Hans Berger, Dr. Alexander Klimakow, Dr. Manfred Mühlberg, Dipl.-Krist. Marion Wienecke
Humboldt-Universität zu Berlin, Sektion Physik, Bereich Kristallographie
Invalidenstr. 110
Berlin
DDR-1040

Cryst. Res. Technol.	23	1988	1	85–95

U. WERNER, J. HEYDENREICH

Institut für Festkörperphysik und Elektronenmikroskopie
der Akademie der Wissenschaften der DDR, Halle (Saale)

Simulation of the Carrier Diffusion Process for the Interpretation of EBIC Profiles

Dedicated to Professor Hermann NEELS on the occasion of his 75th birthday

Based on a Monte Carlo calculation procedure a phenomenological description is developed of the diffusion process of excess minority charge carriers inside a semiconducting target. Using this description and an analytic expression of the generation volume of electron-hole pairs in Si the electron beam induced current (EBIC) obtained by applying a scanning electron microscope can be simulated under consideration of relatively complicated structures of junction semiconductor devices. This allows a better interpretation of EBIC profiles as to the information about the geometric structure of space charge regions built up in junction semiconductor devices. As an example of such an interpretation the verification of the method of CHI and GATOS is presented.

Auf der Grundlage der Monte-Carlo-Methode wird eine phänomenologische Beschreibung des Diffusionsprozesses der Überschuß-Minoritätsladungsträger im Halbleitertarget entwickelt. Unter Anwendung dieser Beschreibung und einer analytischen Beziehung für das Generationsvolumen von Elektronen-Loch-Paaren im Si kann der im Rasterelektronenmikroskop erzeugte EBIC (electron beam induced current) unter Berücksichtigung relativ komplizierter Dotierungsstrukturen im Halbleiter simuliert werden. Dies erlaubt eine fundierte Interpretation von EBIC-Profilen hinsichtlich der Informationen über die geometrische Struktur der von den Dotierungsstrukturen aufgebauten Raumladungszonen. Die Verifizierung der Methode von CHI und GATOS wird als Beispiel einer solchen Interpretation vorgestellt.

1. Introduction

In semiconductor physics and technology the investigation of space charge regions in doped semiconductor crystals — especially in silicon and silicon device structures — gains increasing importance. Working with the scanning electron microscope in the electron beam induced current (EBIC) mode enables the characterization of space charge regions by the interpretation of EBIC profiles. Via simulation of the carrier diffusion process the present paper aims at a realistic description of EBIC produced in silicon device structures in order to attain a reliable characterization of the device space charge regions.

Incident into a semiconducting target the keV electron beam of a scanning electron microscope generates a great number of excess charge carriers (electron-hole pairs) by the interaction with valence electrons. Provided that the diffusing minority charge carriers reach a space charge region (SCR) inside the target the electric field of the SCR separates these carriers from the majority ones. From this process the so-called electron beam induced current (EBIC) arises. Accordingly, this EBIC is determined by the generation, the diffusion and the detection of minority charge carriers. Until recently it had been possible to describe each ot these three processes by drastic simplifications, as for instance: the first by a point source (e.g. FLAT, MILNES; IOANNOU, DIMITRIADIS), the second by treating a vertically arranged and semiinfinitely extended SCR (e.g. BERZ, KUIKEN; SOUKUP), and the third by the assumption of a detection probability of unity (e.g. AKAMATSU et al.; the present paper), respectively. But with the SCR dimsnsions becoming smaller in modern integrated circuits ($\lesssim 1\ \mu m$), at least the first two descriptions are oversimplifications because of the influence of the electron-hole pair generation volume and of the complicated geometric structure of the SCR on EBIC. In order to interpret EBIC profiles under these conditions, calculations must be carried out considering these influences:

The precise consideration of the electron-hole pair generation requires the calculation of the structure of the generation volume, i.e. of the spatial dose distribution of the absorbed energy, caused by the incident electron beam. This calculation was accomplished applying a Monte Carlo (MC) simulation procedure of the electron-target interaction (WERNER, HEYDENREICH) for GaAsP by OELGART and WERNER, and for Si by WERNER et al. In the latter work the authors derived an analytic description of the generation volume of Si — the material investigated here — using the results of the MC simulation and a model of the electron-target interaction. In this model the generation volume is made up of one part that is caused by the spreading penetrating electron beam, and of another part that is caused by electrons which propagate isotropically — like in diffusion spreading — in the depth of the target.

The problem of describing the diffusion of minority carriers consists in the consideration of the boundary conditions resulting from the complicated geometric structures of SCR of modern integrated circuits. If under these requirements analytic solutions of the continuity equation are looked for, each new boundary condition, this means each new structure, requires a new solution, the complexity of which rapidly rises the more complicated the structure becomes. Due to the treatment of the boundary conditions by "if"-statements the MC method is a much more flexible calculus than analytic methods are. The difficulty in solving the diffusion problem of minority carriers via a MC simulation arises from the very long computing time. Therefore, most important for the present work is the creation of a very quick MC simulation procedure for the carrier diffusion treatment.

2. Monte Carlo simulation of the carrier diffusion

In order to derive a sufficiently quick MC simulation the description of the carrier diffusion should be as simple as possible: First of all, the calculation of the carrier trajectories can be confined to two dimensions because only two-dimensional SCR structures are to be dealt with. As, furthermore, solely the phenomenology of the carrier diffusion is of interest single scattering processes can be summarized to isotropic transport impacts. The mean free paths of these transport impacts λ_t and of the carrier absorption λ_a are linked with the carrier diffusion length L, the mainly interesting parameter of the phenomenological description, for two dimensions by

$$L^2 = \tfrac{1}{2} \lambda_t \lambda_a \,. \tag{1}$$

This relation reveals that λ_t and λ_a can be chosen relatively arbitrarily provided that their half product is kept equal to L^2. This means that λ_t can be increased appropriately to decrease the calculation time. Furthermore, the efficiency of the MC simulation can remarkably be enhanced by calculating the probability

$$p(s) = \exp(-s/\lambda_a) \tag{2}$$

for the carrier to cover the way s along its path instead of tracing the actual carrier history. This enables the analytic description of the absorption (recombination) of the carriers on their way throughout the semiconductor, which thus avoids the simulation of the absorption via the MC calculation with its statistic error.

Resulting from the above-suggested scheme the MC simulation contains the following simple core for calculating the carrier paths:

$$\begin{aligned} \vartheta &:= 2\pi R \\ x &:= x + \lambda_t \cdot \sin\vartheta \\ z &:= z + \lambda_t \cdot \cos\vartheta \\ s &:= s + \lambda_t \\ p &:= \exp(-s/\lambda_a)\,, \end{aligned} \tag{3}$$

where x and z denote the lateral and depth positions of the carriers. The "scattering" angles ϑ of the transport impacts are chosen in terms of the random number R equally distributed between 0 and 1. Repeating this formula cycle down to a sufficiently small carrier survival probability $p(s)$ enables a "carrier path" to be simulated. A schematic survey of this simulation and of the determination of the EBIC profile is depicted in Figure 1 for a point source and a semiinfinitely extended SCR vertically arranged to the surface. The carrier starting from the point source (located in segment $N_x = 0$) migrates in the semiconductor and travels through different segments N_x which represent the SCR in different positions. Whenever the carrier enters a segment for the first time (marked by points) the survival probability (2) of the carrier is stored by the array variable $p(N_x)$. In Figure 1b, the histogram resulting from this variable shows the EBIC response function (solid line) gained solely by evaluating the one trajectory shown under the assumption that each carrier is detected that reaches the SCR. Averaging a great number of such profiles helps diminish the strong

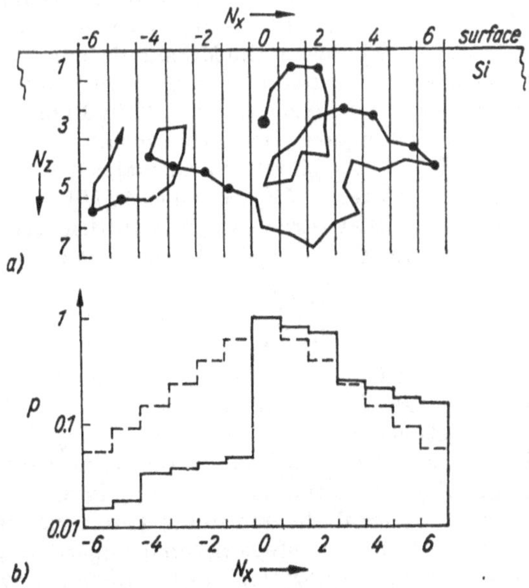

Fig. 1. Schemes of a simulated charge carrier path (a) and of the derived EBIC profile (b). They are sketched for a charge carrier point source (at $N_x = 0$) and a semiinfinitely extended SCR in vertical arrangement to the surface. The different positions of the SCR are indicated by the segment stripes N_x

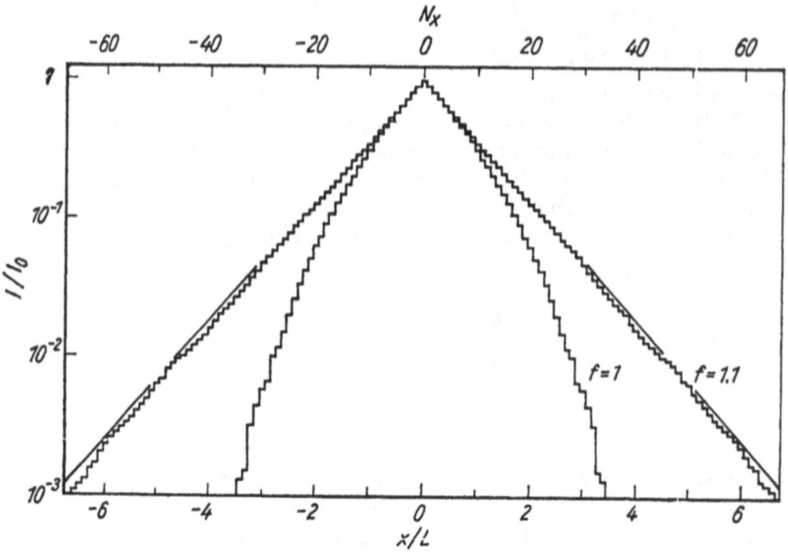

Fig. 2. EBIC profiles calculated after the scheme of Figure 1 with the bases $f = 1$ and 1.1 of the power function in (4). The lines (— —) near the upper histrogram point out the ideal behaviour of the EBIC. Simulation parameters: $S = 0$, $W/L = 0.1$, $\lambda_t/L = \lambda_{t0}/L = 0.1$, $i_0 = 300$ ($f = 1$) and $i_0 = 100$ ($f = 1.1$), $n_0 = 1000$ (S-reduced surface recombination velocity, W – SCR width).

statistic fluctuation. Presupposing that no surface recombination occurs, the averaged EBIC profile would finally display the behaviour as the dashed histogram in Figure 1b reveals.

Since only a transport impact number of $i_0 \lessapprox 500$ can be handled by a microcomputer in reasonable time, during the simulation the mean deviation of the propagating carriers from the point source does not become high enough for a precise description of the real diffusion. Therefore, a correction factor is introduced into relation (1) which is determined by using the MC simulation itself for modelling L precisely. Employing this correction procedure, the MC simulation of the above-presented problem (Fig. 1) yields the lower histogram ($f = 1$) in Figure 2 by calculating $n_0 = 1000$ trajectories per $i_0 = 300$ impacts (as to the remaining parameters, see the figure caption). Here, for $|x|/L > 1$, the simulated normalized EBIC I/I_0 drops drastically below the real one — which reads $I/I_0 = \exp(-|x|/L)$ — since the correction can only model the precise L in that region which is sufficiently traversed by the diffusing carriers. In order to extend this region the mean free paths are modified:

$$\lambda_t := \lambda_{t0} \cdot f^i$$
$$\lambda_a := \lambda_{a0}/f^i \,, \tag{4}$$

where λ_{t0} and λ_{a0} stand for the starting mean free paths, and f^i is a power function with the base $f \gtrapprox 1$ and the actual number of impacts $i = 1 \dots i_0$ as the power. This modification causes λ_t to rise continuously (for a great value of i it increases rapidly); but the product of both mean free paths remains constant, and so does the diffusion length L after relation (1). Accordingly, at the beginning of this path simulation the carrier "explores" the immediate source vicinity by small steps, later on for the more distant vicinity the steps become larger. Accordingly, the above description yields only a rough "exploration" of the distant environment. This is justified by the fact that small SCR details far away from the source are hardly detectable in the EBIC profile. The upper histogram in Figure 2 reveals that using modification (4) the simulation ($n_0 = 1000$, $i_0 = 100$, $f = 1.1$) strongly approaches the real EBIC response function. Diverse tests of the described small simulation program being run on a microcomputer have confirmed the simulation concept presented (for more information about the simulation, see WERNER, EISENSCHMIDT).

3. Calculation of EBIC profiles

An example of treating a more complicated SCR structure is given in Figure 3. The investigation of this structure presupposes the knowledge of the probability of a minority carrier generated at a target point (x, z) to be detected by the SCR. The distribution of this detection probability $p(x, z)$ in the x–z plane is obtained not only by using segment strips as in Figure 1a but by additionally subdividing the z-axis (as indicated in Fig. 1a). This procedure produces a grid in each segment mesh (N_x, N_z) of which a point source is placed. Besides, by employing the segment dimensions $\Delta x = \Delta z = W/13$ (W-SCR width), within this grid the area of the SCR is defined by $-6 \leqq N_x \leqq 6$ if $1 \leqq N_z \leqq 13$ for the vertical SCR part, and by $-6 \leqq N_x \leqq 100$

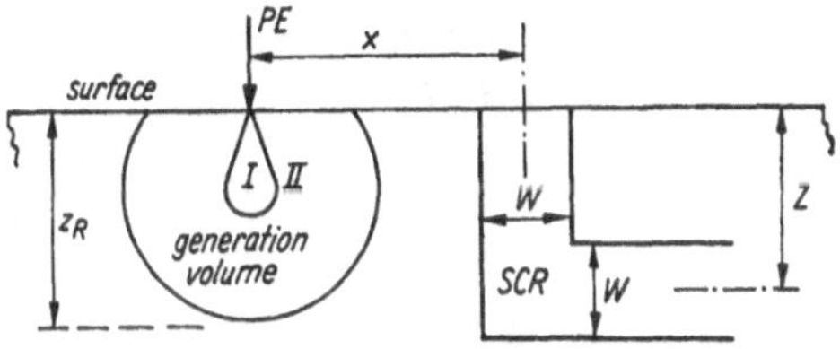

Fig. 3. Outline of the arrangement of the generation volume of electron-hole pairs and the investigated SCR for recording EBIC profiles. Designations: W – SCR width, Z – SCR depth, x – distance between electron beam and the middle of the SCR, Z_R – range of primary electrons (PE), I – penetration and II – diffusion part of the generation volume.

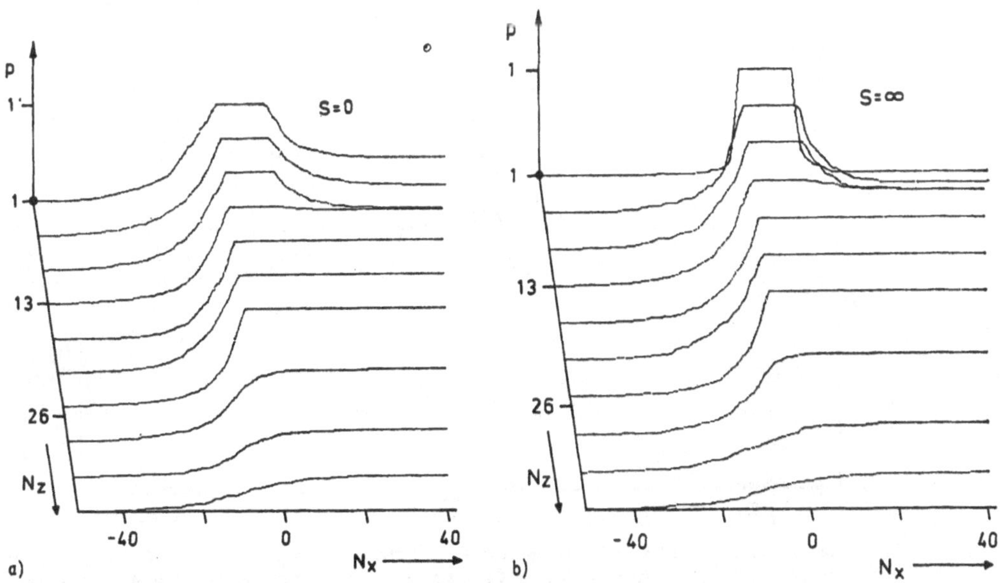

Fig. 4. Distribution of the detection probability p in the N_xN_z – plane for a geometric structure of the SCR as outlined in Figure 3. Simulation parameters: $S = 0$ (a) and $S = \infty$ (b), $W/L = 1.30$, $Z/L = 1.95$, $\lambda_{t0}/L = 0.01$, $\Delta x/L = \Delta z/L = 0.1$, $i_0 = n_0 = 100$

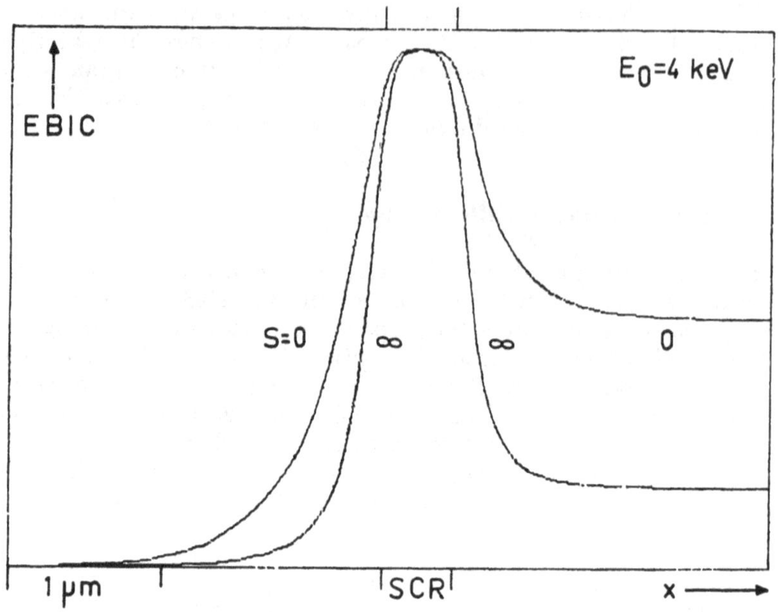

a

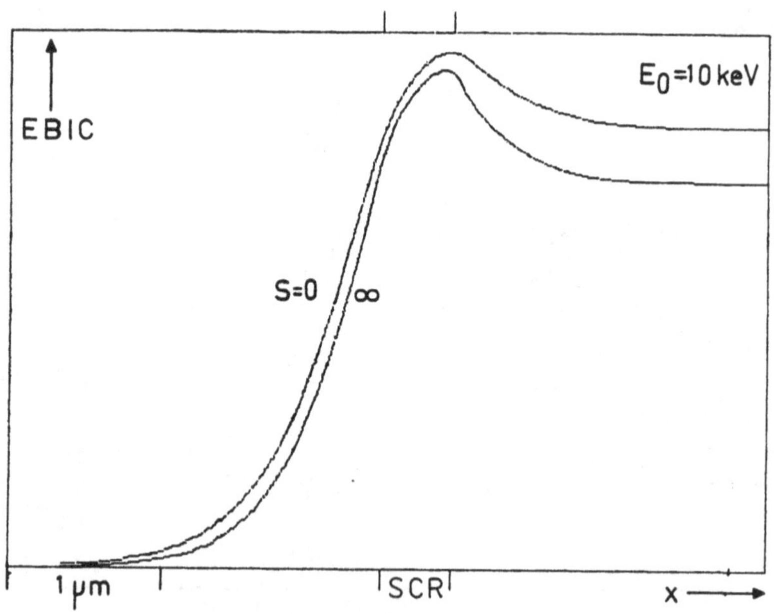

b

Fig. 5. Simulation of EBIC profiles of the SCR structure shown in Figure 3, by using the detection probability distribution of Figure 4. Simulation parameters: E_0 = 4 keV (a) and E_0 = 10 keV (b), W = 0.5 μm, Z = 0.75 μm, L = 0.385 μm, d = 50 nm. The boundaries of the vertical SCR part are marked by dashes on the abscissa

if $14 \leqq N_z \leqq 26$ for the horizontal SCR part. Under the boundary conditions of the SCR structure and the surface the carrier diffusion must subsequently be simulated for ca. 8000 point sources distributed over the grid. The scope of this task can be reduced if the simulations for all point sources located at the same depth are summarized and treated in a way similar to the preceding example. This procedure requires the careful formulation of the boundary conditions by "if"-statements.

The distribution of the detection probability $p(N_x, N_z)$ calculated as depicted above by simulating $n_0 = 100$ paths per point source is shown in Figures 4a and b for reduced surface recombination velocities of $S = 0$ and ∞, respectively.[1]) All curves presented in these figures can also be regarded as "EBIC profiles" produced by moving the point sources at different depths parallel to the surface. In Figure 4a the behaviour of these profiles is governed by the geometric structure of the SCR and its relation to the excess carrier diffusion length L. Choosing the width $W = 1.30 \cdot L$ and depth $Z = 1.95 \cdot L$ of the SCR (see Fig. 3) yields a relation where both the SCR structure and the diffusion process are marked distinctly. In Figure 4b, the high surface recombination causes a profile slope near the surface that is stronger than exponential and which can be traced about one diffusion length from the surface down into the target depth. Moreover, in Figures 4a and b the profile of $N_z = 25$ displays also a slope slightly stronger than exponential, which is due to the absence of the following vertical SCR part infinitely extended into the depth like in the first example.

The real EBIC produced by the incident electron beam is possible to obtain by convoluting the distribution of the carrier detection probability with the generation volume of electron-hole pairs. This requires the projection of the analytic spatial dose distribution (see WERNER et al.) into the plane of the paper and the digital transfer of this projection into a grid similar to that used for the detection probability, depending on the beam diameter d and the incident energy E_0. In changing the distance x between incident electron beam and SCR (see Fig. 3) progressively, the repeated convolution of both grids finally yields EBIC profiles such as those in Figure 5. These profiles calculated for $d = 50$ nm and $E_0 = 4$ and 10 keV consist of two fractions: The first appearing as a peak and the second appearing as a plateau (on the right peak side) are produced by the vertical and horizontal SCR parts, respectively. With increasing incident energy and thus rising penetration depth of the incident electrons the profile peak loses its dominant position as to the profile plateau (cf. Figs 5a and b). As a result, the profile peak also loses its information about location and width of the vertical SCR part (marked by dashes on the abscissa). Thus, the half width of 0.75 μm of the profile peak for $E_0 = 4$ keV and $S = \infty$ (Fig. 5a) enables a rough immediate estimation of the SCR width which is fixed to $W = 0.5$ μm by choosing the diffusion length ($L = 0.385$ μm). But for the higher incident energy of $E_0 = 10$ keV (Fig. 5b) even a rough estimation seems to be impossible. Although in the simulation L is assumed to be identical on both sides of the p–n junction, owing to the horizontal SCR part the profile maximum is shifted with increasing energy (cf. Figs 5a and b), making the experimental determination of the SCR location more uncertain.

4. Verification of the method of CHI and GATOS

CHI and GATOS establish the conception of their diagnostic method by discussing the EBIC production on the basis of a simple spherical generation volume of a constant

---

[1]) The reduction reads $S = s\tau/L$ with s denoting the surface recombination velocity and τ the minority carrier lifetime.

pair creation rate. For a SCR structure, such as in Figure 3, they find that with increasing incident energy E_0 the above-presented profile peak should vanish at a certain energy E_0^*. Since the discussion of the EBIC production combines the electron range of this energy $z_R(E_0^*)$ with the depth Z of the horizontal SCR by setting

$$Z = 0.48 \cdot z_R(E_0^*) \tag{5}$$

the determination of Z becomes possible. Some experiments carried out by CHI and GATOS this formula.

Despite of this success several questions remain open: Thus, errors may be assumed to arise from the simple model of the generation volume used. Furthermore, the discussion considers solely that contribution to the EBIC which is caused by the charge carriers produced in the SCR itself; the diffusion of charge carriers towards the SCR is neglected. The same holds good for the surface recombination. Consequently, it seems to be uncertain whether this very EBIC behaviour is observed for all possible values of L and S. It is therefore necessary to verify the method of CHI and GATOS on a basis more theoretically founded:

In order to perform the first EBIC simulation with parameters which approximately correspond with those of the SCR structure, employed in CHI and GATOS' paper, these parameters were very roughly estimated at $L = 20$ µm, $S = \infty$, $W < < 1$ µm from the measured EBIC profiles published there (Fig. 1b). With the depth of the horizontal SCR of $Z = 3.3$ µm (measured by CHI and GATOS), the SCR width of $W = 0.347$ µm (i.e. $W \ll Z$) and the beam diameter of $d = 100$ nm the simulation finally yields the incident-energy-dependent EBIC profile set of Figure 6a. In spite of the roughly estimated parameters, the simulated profiles and those measured by CHI and GATOS display the same behaviour. With rising incident energy the profile peak clearly diminishes and vanishes almost entirely at $E_0^* = 30$ keV. At this energy, in the location of the vertical SCR (marked by a dash) the slope region of the profile flows strikingly abruptly into a plateau region. With this energy one attains the formula $Z = 0.37 \cdot z_R(E_0^*)$ for the horizontal SCR depth of the electron range

$$z_R = 0.0307 \cdot E_0^{5/3} \tag{6}$$

with E_0 in keV and z_R in µm is used, which was empirically deduced from the employed simulation of the electron scattering in Si and which almost coincides with that applied by CHI and GATOS.

The profile set changes drastically if the surface recombination velocity becomes zero, as it is shown in Figure 6b. Here the diffusion process can largely evolve, whereas in Figure 6a its formation is highly moderated due to the strong influence of the surface recombination ($L \approx 6Z$!). The fully developed diffusion in Figure 6b smoothes the EBIC profiles out and cuts the information down. Only the evaluation of the numerical results (exhibiting a weak peak) allows the estimation of the energy (with the peak vanishing) at about $E_0^* = 25$ keV, from which $Z = 0.50 \cdot z_R(E_0^*)$ can be inferred.

If the diffusion length L is shortened to 1 µm with the remaining parameters, however, being maintained, Figures 6a and b are converted into Figures 6c and d, respectively. Because of $L = 0.3 \cdot Z$, in the latter the surface recombination solely has a slight effect on both the profile sets, so that they are similar. On the other hand, the influence of the structure of the generation volume has increased. With the peak and the right tail flowing into the plateau, particularly the profiles for $E_0 = 28$ keV (Fig. 6c) and $E_0 = 29$ keV (Fig. 6d) show the influence of the penetration as well as of the diffusion part of the generation volume, respectively (cf. WERNER et al.). In searching for the energy E_0^* in both figures one cannot find a profile where the peak vanishes. However, carefully studying Figure 6a reveals a similar fact; here, for

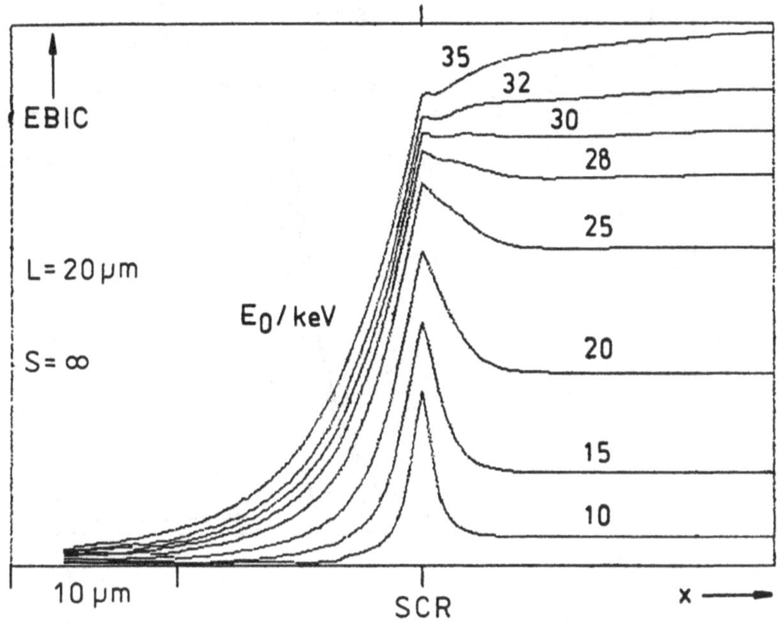

EBIC
L = 20 μm
S = ∞
E0 / keV
35
32
30
28
25
20
15
10
10 μm
SCR
x

Fig. 6a

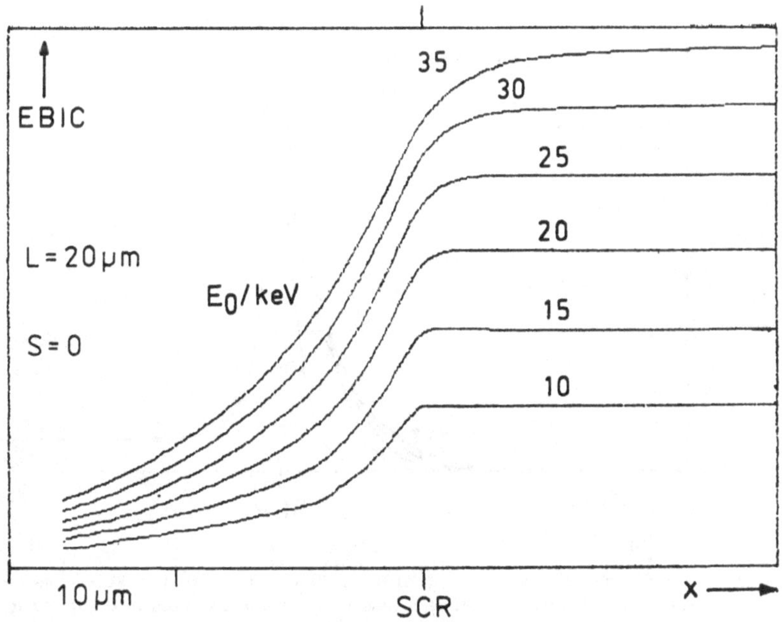

EBIC
L = 20 μm
S = 0
E0 / keV
35
30
25
20
15
10
10 μm
SCR
x

Fig. 6b

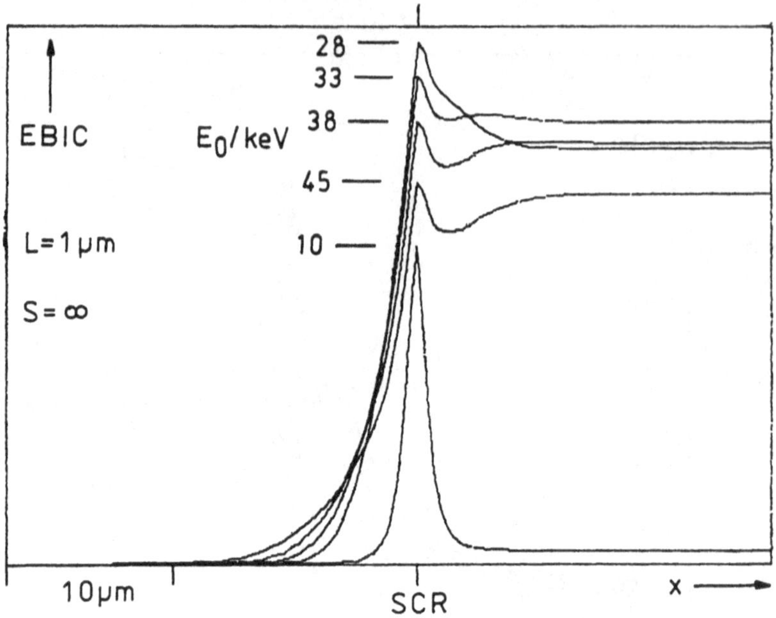

Fig. 6 c

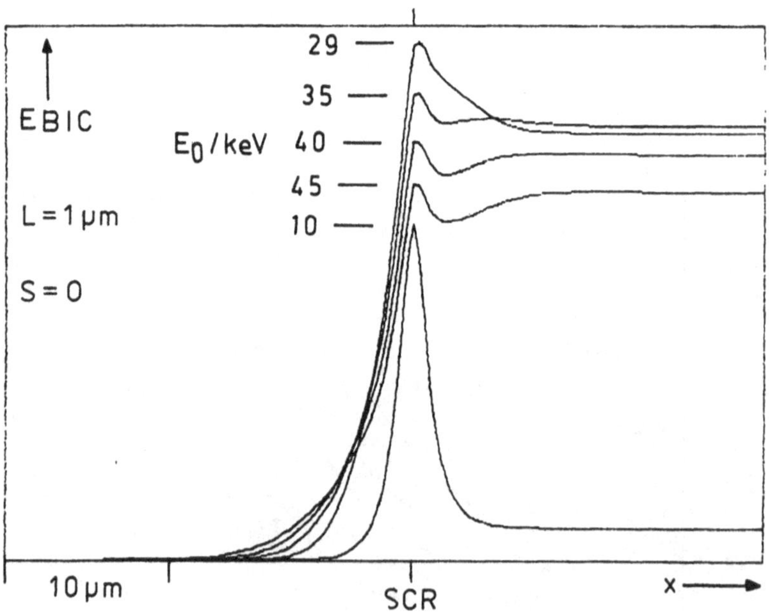

Fig. 6d

Fig. 6. Energy dependence of EBIC profiles resulting from a SCR structure similar to that shown in Figure 3 for diffusion lengths of 20 µm and 1 µm in (a, b) and (c, d), respectively, and for reduced surface recombination velocities of ∞ and 0 in (a, c) and (b, d), respectively. Further simulation parameters: $W = 0.347$ µm, $Z = 3.3$ µm, $d = 100$ nm. The position of the vertical SCR part is indicated by a dash on the abscissa

energies $E_0 \geqq 30$ keV the profiles also display a rest of a peak. Therefore a more general criterion of determining E_0^* may be the most extended plateau, i.e. the plateau which most closely approaches the position of the vertical SCR. With this criterion, the profiles of Figures 6c and d with $E_0^* = 33$ keV and $E_0^* = 35$ keV lead to $Z = 0.32 \cdot z_R(E_0^*)$ and $Z = 0.29 \cdot z_R(E_0^*)$, respectively.

Summarizing all the formulas obtained by combining relatively extreme values of the parameters L and S yields the universal formula

$$Z = (0.4 \pm 0.1) \cdot z_R(E_0^*) . \tag{7}$$

Provided that the most extended plateau can distinctly be observed experimentally, using this formula enables the depth of the horizontal SCR to be estimated without knowledge of the diffusion length and the surface recombination velocity. This verification of CHI and GATOS' method illustrates how the peculiarity of the developed program to simulate an idealized measurement can be applied to establish simple formulas which disclose the influence of SCR structure parameters on EBIC.

References

AKAMATSU, B., HENOC, J., HENOC, P.: J. Appl. Physics **52**, 7245 (1981)
BERZ, F. KUIKEN, H. K.: Solid-State Electronics **19**, 437 (1976)
CHI, J.-Y., GATOS, H. C. IEEE Trans. Electron. Devices **ED-24**, 1366 (1977)
FLAT, A., MILNES, A. G.: Int. J. Electronics **44**, 629 (1978)
IOANNOU, D. E., DIMITRIADIS, C. A.: IEEE Trans. Electron Devices **ED-29**, 445 (1982)
OELGART, G., WERNER, U.: phys. stat. sol. (a) **85**, 205 (1984)
SOUKUP, R. J.: Appl. Phys. Communications **2** (**3**), 143 (1982–83)
WERNER, U., EISENSCHMIDT, Ch.: Research Report, Inst. für Festkörperphysik und Elektronenmikroskopie, Halle (Saale) 1987
WERNER, U., HEYDENREICH, J.: Ultramicroscopy **15**, 17 (1984)
WERNER, U., KOCH, F., OELGART, G.: J. Phys. D: Appl. Phys., to be published

(Received July 17, 1987)

Authors' address:

Dr. U. WERNER, Prof. Dr. J. HEYDENREICH
Akademie der Wissenschaften der DDR
Institut für Festkörperphysik und Elektronenmikroskopie
Weinberg 2
Halle (Saale)
DDR-4050

Cryst. Res. Technol.	23	1988	1	97–102

H. NEUMANN

Sektion Physik, Karl-Marx-Universität Leipzig

Microhardness Scaling and Bulk Modulus-Microhardness Relationship in $A^{II}B^{IV}C_2^V$ Chalcopyrite Compounds

Dedicated to Professor Hermann NEELS on the occasion of his 75th birthday

An analysis of experimental microhardness data shows that the microhardness of the $A^{II}B^{IV}C_2^V$ compounds can be described by a simple scaling law which depends only on the unit cell volume and the melting temperature. On the basis of this result a simple bulk modulus – microhardness relationship is proposed and used to estimate the bulk moduli of the compounds. The results are compared with existing literature data.

Eine Analyse experimenteller Mikrohärtedaten zeigt, daß die Mikrohärte der $A^{II}B^{IV}C_2^V$-Verbindungen durch ein einfaches Skalierungsgesetz beschrieben werden kann, das nur vom Elementarzellenvolumen und der Schmelztemperatur abhängt. Auf der Grundlage dieses Resultats wird eine einfache Beziehung zwischen Kompressionsmodul und Mikrohärte vorgeschlagen und zur Berechnung der Kompressionsmoduln der Verbindungen ausgenutzt. Die Ergebnisse werden mit vorhandenen Literaturdaten verglichen.

1. Introduction

Although the properties of the $A^{II}B^{IV}C_2^V$ chalcopyrite semiconductors have been extensively investigated and some of these compounds have attracted attention for practical applications (BOUDRIOT et al.; GORUYNOVA, VALOV; PROCHUKHAN, RUD; SHAY, WERNICK) the knowledge of their elastic properties is rather incomplete. Experimental data are available only for two compounds. For $CdGeAs_2$ a complete determination of the elastic constants including their temperature and pressure dependence has been reported (HAILING et al.). In the case of $CdSiP_2$ the bulk modulus has been deduced from X-ray measurements at high pressures (BETTINI, HOLZAPFEL). Thus it is not surprising that various attempts have been made to estimate the bulk modulus of these compounds from other material parameters which are better investigated (DEUS, SCHNEIDER; OSHCHERIN). Recently, also direct calculations of the elastic constants on the basis of a rigid ion model have been reported (POPLAVNOI). However, if these estimated bulk moduli are compared with the experimental ones one finds that none of these models gives satisfactory agreement with experiment (NEUMANN 1986). In a recent study of the scaling laws for the bulk modulus and the microhardness of the tetrahedrally coordinated elemental and binary semiconductors it has been established that there is a simple relationship between the bulk modulus and the microhardness that depends only on the melting temperature, the unit cell volume and the spectroscopically defined bond ionicity (NEUMANN 1987a, b). In the present work it is shown that an analogous relation exists for the ternary chalcopyrite compounds which can be successfully employed to estimate the bulk modulus from experimental microhardness data that are available for most of the $A^{II}B^{IV}C_2^V$ semiconductors.

2. Microhardness scaling law

In analysing the trends in the microhardness of the tetrahedrally bonded elemental and binary semiconductors (NEUMANN 1987b) it has been found that the microhardness follows the simple relation

$$H = hT_mV_0^{-n} \tag{1}$$

where T_m is the melting temperature, V_0 the unit cell volume and h a function of the bond ionicity f_i given by

$$h = h_0(1 - h_1f_i) \,. \tag{2}$$

The constants h_0, h_1 and n have the same values for all these compounds. To verify the validity of relations (1) and (2) for the ternary $A^{II}B^{IV}C_2^V$ compounds we are confronted with the problem that we have two types of bonds and therefore also two different individual bond ionicities (LEVINE). In particular, it is an open question how the function h must be defined in terms of the two bond ionicities. Fortunately, the individual bond ionicities differ only slightly from compound to compound. Furthermore, if we evaluate the average bond ionicities of the $A^{II}B^{IV}C_2^V$ compounds according to NEUMANN (1983) using the same input parameters as LEVINE for the susceptibilities and the lattice parameters reported by GORYUNOVA, VALOV, and by VAIPOLIN we obtain the f_i values given in Table 1. We see that except in $ZnSnAs_2$ the average bond ionicities vary in the rather narrow range $f_i = 0.33 \pm 0.05$. Then, assuming that the average bond ionicity can be used to approximately estimate the function h with relation (2) and taking into account the uncertainties in the susceptibilities for the $A^{II}B^{IV}C_2^V$ compounds used in the calculations (LEVINE) as well as the usual error of about 10% in microhardness determinations (WOLFF et al.) it seems to be justified to suppose that the function h has nearly the same value for all compounds in analysing the microhardness data.

The microhardness of the $A^{II}B^{IV}C_2^V$ compounds has been studied by various authors (ABRAHAMS, HSU; BORSHCHEVSKI et al. 1967, 1971; GORYUNOVA, VALOV; OSMANOV et al.). From an analysis of the rather detailed results of BORSHCHEVSKI et al. (1971) and of the data reviewed by GORYUNOVA, VALOV we estimated the microhardness values and their error limits given in Table 2. Using these hardness values, the melting temperatures T_m from GORYUNOVA, VALOV and the lattice parameters a and c from GORYUNOVA, VALOV and from VAIPOLIN we have calculated H/T_m and $V_0 = a^2c$. Figure 1 shows a double-logarithmic plot of H/T_m against V_0. We see that indeed relation (1) with a constant value for the function h can be used to describe the microhardness of the $A^{II}B^{IV}C_2^V$ compounds. The fit of relation (1) to the data points of Figure 1 yields $h = 1.399 \cdot 10^{-3}$ and $n = 1.436$ if H is inserted in GPa, T_m in K and

Table 1
Average bond ionicities f_i of the $A^{II}B^{IV}C_2^V$ compounds

Compound	f_i	Compound	f_i
$ZnSiP_2$	0.30	$CdSiP_2$	0.36
$ZnGeP_2$	0.30	$CdGeP_2$	0.36
$ZnSnP_2$	0.32	$CdSnP_2$	0.38
$ZnSiAs_2$	0.36	$CdSiAs_2$	0.38
$ZnGeAs_2$	0.29	$CdGeAs_2$	0.33
$ZnSnAs_2$	0.25	$CdSnAs_2$	0.29

Table 2
Experimental and calculated microhardness (all values in GPa) of the $A^{II}B^{IV}C_2^{V}$ compounds

Compound	H_{exp}	H_{calc}		
		this work	GARBATO, RUCCI	SAMANTA et al.
$ZnSiP_2$	12.3 ± 1.0	12.2	8.8	10.4
$ZnGeP_2$	9.8 ± 0.5	9.3	6.2	7.6
$ZnSnP_2$	7.0 ± 0.4	7.3	5.2	4.9
$ZnSiAs_2$	8.8 ± 0.5	8.9	8.0	8.7
$ZnGeAs_2$	6.6 ± 0.5	6.9	6.2	6.7
$ZnSnAs_2$	5.4 ± 0.2	5.4	4.2	4.4
$CdSiP_2$	9.3 ± 0.5	9.3	7.2	7.5
$CdGeP_2$	6.7 ± 0.4	6.6	4.6	5.1
$CdSnP_2$	5.3 ± 0.4	5.3	2.5	2.7
$CdSiAs_2$	6.1 ± 0.5	6.4	6.0	6.3
$CdGeAs_2$	5.2 ± 0.4	5.0	4.6	4.8
$CdSnAs_2$	4.0 ± 0.4	3.9	3.0	3.1

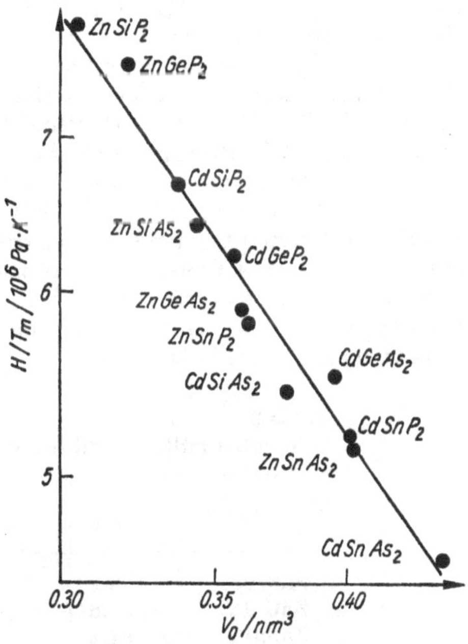

Fig. 1. Double-logarithmic plot of H/T_m against V_0

V_0 in nm^3. The microhardness values calculated with (1) and these parameters are given in Table 2, too.

For comparison, we have included in Table 2 also the microhardness estimates reported by GARLATO, RUCCI, and by SAMANTA et al. Obviously, the results of these authors agree with experiment only for $ZnSiAs_2$, $ZnGeAs_2$, $CdSiAs_2$ and $CdGeAs_2$. For the remaining compounds the microhardness values estimated by these authors are evidently too low.

3. Calculation of the bulk modulus

In analysing the trends in the bulk moduli of the elemental and binary semiconductors with tetrahedral coordination (NEUMANN 1987a) it has been found that the bulk modulus is only a function of the unit cell volume and the spectroscopically defined bond ionicity and can be written in the simple form

$$B = bV_0^{-m} \tag{3}$$

with

$$b = b_0(1 - b_1 f_i) \tag{4}$$

and the same values for the constants b_0, b_1, and m for all compounds. Relation (3) has been found to be valid also in the ternary $A^{II}B^{IV}C_2^V$ chalcopyrite compounds (NEUMANN 1986). Then, if we assume that a relation of the type (3) describes also the bulk moduli of the $A^{II}B^{IV}C_2^V$ compounds we can combine relations (1) and (3) and obtain the simple relationship

$$B = cV_0^{-k}H/T_m \tag{5}$$

between the bulk modulus and the microhardness with the constants c and k to be determined by comparison with experiment. In analogy to the function h (relation (2)) it is certainly justified to suppose that the function b given by relation (4) has the same constant value for all $A^{II}B^{IV}C_2^V$ compounds. Then, using the experimental bulk modulus values of 69.7 GPa for $CdGeAs_2$ (HAILING et al.) and of 97 GPa for $CdSiP_2$ (BETTINI, HOLZAPFEL) we can determine c and k. We find $c = 7.18 \cdot 10^3$ and $k = 0.6475$ if V_0 is inserted in nm^3 and T_m in K. The bulk moduli calculated with relation (5) with these constants and the calculated microhardnesses from Table 2 are compiled in Table 3. For comparison, the B values estimated from the Debye temperature (DEUS, SCHNEIDER) and the surface energy (OSCHERIN) as well a the bulk moduli calculated from the elastic constants reported by POPLAVNOI are also given. We see that both the experimental bulk moduli for $CdGeAs_2$ and $CdSiP_2$ as well as the B values estimated from the microhardness are distinctly larger than the estimates of the other authors. Considering the situation in the case of $CdGeAs_2$ and $CdSiP_2$ this trend is certainly correct. However, in order to get more evidence regarding the reliability of the absolute values of the bulk moduli estimated in the present work further experimental bulk modulus determinations for other compounds are required.

Table 3
Calculated bulk moduli (all values in GPa) of the $A^{II}B^{IV}C_2^V$ compounds

Compound	this work	DEUS, SCHNEIDER	OSHCHERIN	POPLAVNOI
$ZnSiP_2$	120.3	73.2	79.0	66.9
$ZnGeP_2$	107.6	86.0	73.6	71.2
$ZnSnP_2$	84.2	—	67.5	57.7
$ZnSiAs_2$	93.4	69.6	68.0	53.2
$ZnGeAs_2$	85.7	53.8	67.0	—
$ZnSnAs_2$	67.0	56.6	55.8	—
$CdSiP_2$	97.0	71.9	48.0	60.9
$CdGeP_2$	86.4	67.4	47.0	60.4
$CdSnP_2$	67.2	55.9	52.5	50.2
$CdSiAs_2$	77.0	—	54.0	—
$CdGeAs_2$	69.7	47.7	41.3	—
$CdSnAs_2$	54.8	49.9	37.1	—

In conclusion, some remarks concerning the constants m, b_0 and b_1 in the relations (3) and (4). For the exponent m we find from the experimentally determined values for the exponents n and k in (1) and (5), respectively, the value $m = n + k = 2.084$ which is very close to that of $m = 2.1035$ deduced from the bulk modulus — volume relationship for the $A^IB^{III}C_2^{VI}$ compounds (NEUMANN 1986). This result confirms the general trend that the exponent m in relation (3) is expected to have the same value for isostructural compounds with the same molecular formula and the same chemical bonding properties. For the constant b we find $b = hc = 10.04$ which is larger than the value of $b = 7.333$ determined for the $A^IB^{III}C_2^{VI}$ compounds (NEUMANN 1986). Considering relation (4) for b this result is not unexpected since the mean value for the average bond ionicity of $f_i = 0.33$ is lower than that of $f_i = 0.72$ in the $A^IB^{III}C_2^{VI}$ compounds (NEUMANN 1983, 1986). If we assume that relation (4) in its simple form remains valid for the ternary chalcopyrite compounds and can be used to evaluate by inserting the average bond ionicity f_i we can determine the parameters b_0 and b_1. The result is $b_0 = 12.33$ and $b_1 = 0.563$. The value of b_0 found here for the chalcopyrite semiconductors is very close to that of $b_0 = 10.53$ found for the elemental and binary semiconductors (NEUMANN 1987a). In the case of the constant b_1 the value for the chalcopyrite compounds is about twice as large as that of $b_1 = 0.25$ for the tetrahedrally bonded binary $A^{III}B^V$ and $A^{II}B^{VI}$ semiconductors (NEUMANN 1987a). Obviously, detailed theoretical model studies of the elastic constants in the chalcopyrite compounds and their dependence on the individual or average bond ionicities are required in order to elucidate the reason for the large difference in the coefficients b_1 for binary and ternary compounds.

References

ABRAHAMS, S. C., HSU, F. S. L.: J. Chem. Phys. **63**, 1162 (1975)

BETTINI, M., HOLZAPFEL, W. B.: Solid State Commun. **16**, 27 (1975)

BORSHCHEVSKI, A. S., GORYUNOVA, N. A., KESAMANLY, F. P., NASLEDOV, D. N.: phys. stat. sol. **21**, 9 (1967)

BORSHCHEVSKI, A. S., GORYUNOVA, N. A., TAKHTAREVA, N. K.: in: Complex Semiconductors and Their Physical Properties (ed. S. I. RADAUTSAN), Kishinev 1971, p. 47

BOUDRIOT, H., BUHRIG, E., OETTEL, H., SCHNEIDER, H. G.: in: Verbindungshalbleiter (eds. K. UNGER and H. G. SCHNEIDER), Leipzig 1986, p. 339

DEUS, P., SCHNEIDER, H. A.: Cryst. Res. Technol. **20**, 867 (1985)

GARBATO, L., RUCCI, A.: Phil. Mag. **35**, 1685 (1977)

GORYUNOVA, N. A., VALOV, Yu. A.: $A^2B^4C_2^5$ Semiconductors, Moscow 1974

HAILING, T., SAUNDERS, G. A., LAMBSON, A. W., FEIGELSON, R. S.: J. Phys. C **15**, 1399 (1982)

LEVINE, B. F.: J. Chem. Phys. **59**, 1463 (1973)

NEUMANN, H.: Cryst. Res. Technol. **18**, 1391 (1983)

NEUMANN, H.: phys. stat. sol. (a) **96**, K121 (1986)

NEUMANN, H.: Cryst. Res. Technol. **22**, 99 (1987a)

NEUMANN, H.: Cryst. Res. Technol. **22**, 271 (1987b)

OSHCHERIN, B. N.: phys. stat. sol. (a) **51**, K175 (1979)

OSMANOV, E. O., RUD, Yu. V., STRYALKOVSKI, M. E.: phys. stat. sol. **26**, 85 (1968)

POPLAVNOI, A. S.: Izv. Vyssh. Ucheb. Zav. SSSR, Fizika, **29**(8), 5 (1986)

PROÇHUKHAN, V. D., RUD, Yu. V.: Fiz. Tekh. Poluprov. **12**, 209 (1978)

SAMANTA, L., GHOSH, D. K., BHAR, G. C.: Chem. Phys. **79**, 361 (1983)

SHAY, J. L., WERNICK, J. H.: Ternary Chalcopyrite Compounds: Growth, Electronic Properties, and Applications, Oxford 1975

VAIPOLIN, A. A.: Cryst. Res. Technol. **18**, 415 (1983)
WOLFF, G. A., TOMAN, L., FIELD, N. J., CLARK, J. C.: in :Halbleiter und Phosphore (eds. M. SCHÖN and H. WELKER), Braunschweig 1958, p. 463

(Received May 22, 1987)

Author's address:

Dr. H. NEUMANN
Sektion Physik, Karl-Marx-Universität
Linnéstraße 5
Leipzig 5
DDR-7010

Cryst. Res. Technol.	23	1988	1	103–106

E. NOWAK, H. NEUMANN, M. S. OMAR

Sektion Chemie, Wissenschaftsbereich Kristallographie, and Sektion Physik, Karl-Marx-Universität Leipzig, and School of Physics, University of Bath

Heat Capacity of $Ag_6Ge_{10}P_{12}$ from 180 to 550 K

Dedicated to Professor Hermann NEELS on the occasion of his 75th birthday

The molar heat capacity at constant pressure of $Ag_6Ge_{10}P_{12}$ is measured in the temperature range from 180 to 550 K and the standard molar enthalpies and entropies relative to 298.15 K are calculated on the basis of these data. From a comparison of the temperature variation of the apparent Debye temperatures in $CuGe_2P_3$ and $Ag_6Ge_{10}P_{12}$ it is concluded that lattice anharmonicity is more pronounced in $Ag_6Ge_{10}P_{12}$ than in $CuGe_2P_3$.

Die molare Wärmekapazität bei konstantem Druck wurde für $Ag_6Ge_{10}P_{12}$ im Temperaturbereich von 180 bis 550 K gemessen und daraus die molaren Standardenthalpien und Standardentropien relativ zu 298.15 K berechnet. Aus einem Vergleich der Temperaturabhängigkeit der scheinbaren Debye-Temperaturen in $CuGe_2P_3$ und $Ag_6Ge_{10}P_{12}$ wird geschlossen, daß die Gitteranharmonizität in $Ag_6Ge_{10}P_{12}$ stärker ausgeprägt ist als in $CuGe_2P_3$.

1. Introduction

The ternary compound $Ag_6Ge_{10}P_{12}$ belongs to a novel type of cluster compounds with an interesting combination of structural effects (VON SCHNERING, HÄUSLER) which are expected to influence many of its physical properties. Recent experimental investigations of this compound are related to the elastic constants and their pressure dependence, the infrared lattice vibrations, the optical properties in the range of the fundamental absorption edge and the electrical characteristics (KÖHLER et al.; MACDONALD et al.; NEUMANN et al. 1985b). Electronic structure calculations and detailed considerations concerning the bonding relationships in this type of compounds have also been reported (BULLETT, DAWSON; BULLETT, WITCHLOW). In the present work we report the first experimental study of the thermal properties of $Ag_6Ge_{10}P_{12}$. The heat capacity at constant pressure is measured in the temperature range from 180 to 550 K, and standard enthalpies and entropies relative to 298.15 K are calculated on the basis of these data. Furthermore, the problem of lattice anharmonicity is shortly discussed.

2. Experimental

The samples used in the experiments were prepared from single crystals grown by the Bridgman technique (NEUMANN et al. 1985b). The compound itself was synthesized in closed ampoules at a temperature of about 1300 K from stoichiometric mixtures of the elements with some excess of phosphorus added to the melt as described by VON SCHNERING and HÄUSLER. The heat capacity at constant pressure was measured in the temperature range from 180 to 550 K using a Setaram heat flow calorimeter model DSC 111. Details of the sample preparation and the measurement procedure including the calibration of the calorimeter have been described elsewhere (NEUMANN et al. 1985a) and will not be repeated here. The anticipated error in the measured heat capacities is about 1%.

3. Results and discussion

The measured temperature dependence of the molar heat capacity at constant pressure C_p of $Ag_6Ge_{10}P_{12}$ is represented in Figure 1. To determine smoothed values of C_p and to evaluate the standard thermodynamic quantities we tried to describe the experimental $C_p(T)$ data by various polynomials. The relation

$$C_p(T) = a - b/T^2 + cT - dT^2 \quad (1)$$

we found to give a satisfactory fit to the experimental result over the whole temperature range within the error limits for the measured C_p values. The coefficients of relation (1) were determined by a least-squares fit of (1) to the data points and were found to be $a = 640 \cdot 3\ \mathrm{Jmol^{-1}\,K^{-1}}$, $b = 1.0992 \cdot 10^7\ \mathrm{JKmol^{-1}}$, $c = 4.7577 \cdot 10^{-1}\ \mathrm{Jmol^{-1}\,K^{-2}}$ and $d = 5.6097 \cdot 10^{-4}\ \mathrm{Jmol^{-1}\,K^{-3}}$. The resulting smoothed heat capacities for selected temperatures between 200 and 550 K and the standard enthalpies and entropies relative to 298.15 K calculated by using relation (1) for C_p are compiled in Table 1.

Now, let us consider the problem of lattice anharmonicity in $Ag_6Ge_{10}P_{12}$. Recently, on the basis of an analysis of the third-order elastic constants and of the mean acoustic mode Grüneisen parameters of GaP, $CuGe_2P_3$ and $Ag_6Ge_{10}P_{12}$ it has been argued that the lattice anharmonicity of these compounds increases continuously going from GaP to $CuGe_2P_3$ and $Ag_6Ge_{10}P_{12}$ (MACDONALD et al.). For GaP and $CuGe_2P_3$ this trend

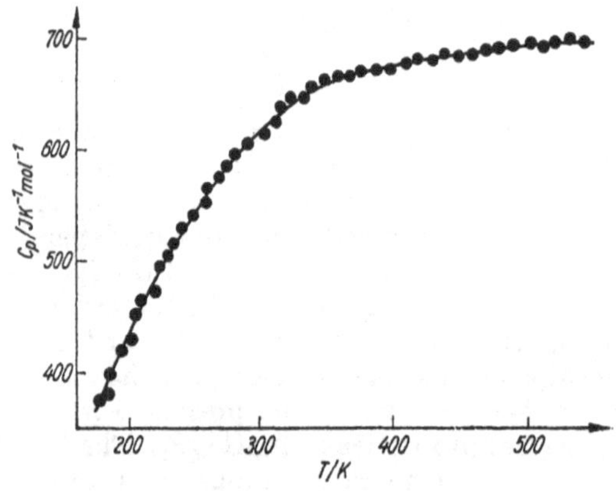

Fig. 1. Temperature dependence of the molar heat capacity at constant pressure of $Ag_6Ge_{10}P_{12}$

Table 1
Standard thermodynamic functions of $Ag_6Ge_{10}P_{12}$ at $p = 101325$ Pa

T K	C J/K mol	$H_m(T) - H_m(298.15)$ kJ/mol	$S_m(T) - S_m(298.15)$ J/K mol
200	438.2	−52.92	−213.1
250	548.3	−27.97	−102.2
300	610.4	1.128	3.77
350	648.3	32.67	100.9
400	672.1	65.73	189.2
450	686.5	99.72	269.3
500	693.9	134.3	342.0
550	696.0	169.0	408.3

has been confirmed by comparing the Grüneisen functions and the thermal expansion coefficients of these compounds at elevated temperatures (NEUMANN et al. 1987b). Unfortunately, the thermal expansion behaviour of $Ag_6Ge_{10}P_{12}$ has not been studied so far, and therefore we are also unable to calculate the Grüneisen function of this compound. Thus, at the present state of knowledge we can only try to get some information about lattice anharmonicity in $Ag_6Ge_{10}P_{12}$ from our heat capacity data.

In order to estimate the anharmonic contribution to the heat capacity we need the Debye temperature θ_∞ in the high-temperature limit $T \to \infty$ which can be deduced from the apparent Debye temperature in a temperature range where anharmonicity effects can be neglected (BARRON et al.). For this reason, we have calculated the apparent Debye temperature $\theta_D(T)$ of $Ag_6Ge_{10}P_{12}$ from the heat capacity data of Figure 1. The resulting temperature variation of θ_D is shown in Figure 2. We see that in the temperature range $T \geqq 180$ K considered in the present work the apparent Debye temperature decreases continuously with increasing temperature. Since the decrease in θ_D can be ascribed to lattice anharmonicity effects (BARRON et al.) it follows from Figure 2 that we have already pronounced lattice anharmonicity effects on C_p over the whole temperature range $T \geqq 180$ K. Thus, we are unable to estimate θ_∞ and, therefore, also the anharmonic contribution to C_p on the basis of our data. However, some qualitative conclusions regarding the degree of lattice anharmonicity in $CuGe_2P_3$ and $Ag_6Ge_{10}P_{12}$ can be made by comparing the temperature dependences of θ_D in these compounds. In the case of $CuGe_2P_3$ it has been found that θ_D increases with temperature up to about 200 K and only then the decrease in θ_D sets in (NEUMANN et al. 1987a). Since the high-temperature limit of the Debye temperature of $CuGe_2P_3$ has been estimated to be about $\theta_\infty = 710$ K it follows from the temperature dependence of θ_D that in this compound the influence of lattice anharmonicity on the heat capacity becomes important at reduced temperatures in the range $T/\theta_\infty \geqq 0.28$. For $Ag_6Ge_{10}P_{12}$ it is obvious from Figure 2 that θ_∞ is larger than 660 K. Moreover, comparing the vibrational characteristics of $CuGe_2P_3$ (RIEDE et al.) and $Ag_6Ge_{10}P_{12}$ (KÖHLER et al.; NEUMANN et al. 1985b) with each other it cannot be excluded that the high-temperature limit of the Debye temperature in $Ag_6Ge_{10}P_{12}$ is even larger than that in $CuGe_2P_3$ for two reasons. Firstly, the vibrational spectrum of $Ag_6Ge_{10}P_{12}$ extends to frequencies of about 400 cm^{-1} with at least five infrared active modes having frequencies above 300 cm^{-1} (KÖHLER et al.; NEUMANN et al. 1985b). From the group theoretical analysis of the lattice vibrations in $Ag_6Ge_{10}P_{12}$ it follows that there should be a nearly equal number of Raman active optical modes in this frequency range (NEUMANN et al. 1985b). This situation is very similar to that in Cu·

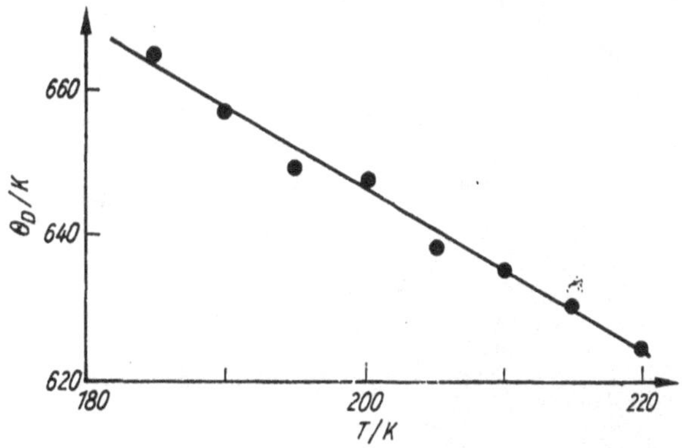

Fig. 2. Temperature dependence of the apparent Debye temperature in $Ag_6Ge_{10}P_{12}$

Ge_2P_3 where we have nine optical modes, all with frequencies near 375 cm^{-1} (NEUMANN et al. 1987a; RIEDE et al.). Secondly, besides the above mentioned high-energy optical modes there is a large number of optical modes at lower energies in $Ag_6Ge_{10}P_{12}$ since the primitive unit cell of this compound contains one formula unit which gives rise to a total of 81 optical phonon branches (NEUMANN et al. 1985b). Thus, since the situation regarding the high-energy optical modes is very similar in $CuGe_2P_3$ and $Ag_6Ge_{10}P_{12}$ and since the modes at lower energies additionally present in $Ag_6Ge_{10} \cdot P_{12}$ also contribute to the phonon density of states it is certainly justified to suppose that the high-temperature limit of the Debye temperature in $Ag_6Ge_{10}P_{12}$ is at least equal to that in $CuGe_2P_3$. Then, it follows from Figure 2 that lattice anharmonicity effects on the heat capacity set in at lower reduced remperatures T/θ_∞ than in $CuGe_2 \cdot P_3$. This result indicates that lattice anharmonicity is more pronounced in $Ag_6Ge_{10}P_{12}$ than in $CuGe_2P_3$ which confirms the conclusions made by MACDONALD et al. on the basis of an analysis of the elastic properties of both compounds.

References

BARRON, T. H. K., BERG, W. T., MORRISON, J. A.: Proc. Roy. Soc. (London) A **242**, 478 (1957)

BULLETT, D. W., DAWSON, W. G.: J. Phys. C **19**, 5837 (1986)

BULLETT, D. W., WITCHLOW, G. P.: Phys. Rev. B **33**, 2429 (1986)

KÖHLER, K., QUEISSER, H. J., HÖNLE, W., VON SCHNERING, H. G., BÖHM, M. C.: Phys. Rev. B **31**, 6514 (1985)

MACDONALD, J. E., SAUNDERS, G. A., YOGURTCU, Y. K., HÖNLE, W.: J. Phys. Chem. Solids **46**, 951 (1985)

NEUMANN, H., KÜHN, G., MÖLLER, W.: Cryst. Res. Technol. **20**, 1225 (1985a)

NEUMANN, H., NOWAK, E., OMAR, M. S.: Cryst. Res. Technol. **22**, Nr. 11 (1987a)

NEUMANN, H., NOWAK, E., OMAR, M. S., PAMPLIN, B. R.: J. Less-Common Metals **132**, L 13 (1987b)

NEUMANN, H., RIEDE, V., SOBOTTA, H., OMAR, M. S., PAMPLIN, B. R.: Solid State Commun. **53**, 155 (1985b)

RIEDE, V., NEUMANN, H., SOBOTTA, H., KISSINGER, W., OMAR, M. S., PAMPLIN, B. R.: phys. stat. sol. (b) **123**, K119 (1984)

VON SCHNERING, H. G., HÄUSLER, K.-G.: Rev. Chim. Mineral. **13**, 71 (1976)

(Received June 12, 1987)

Authors' addresses:

Dr. E. NOWAK
Sektion Chemie (WB Kristallographie)
Karl-Marx-Universität
Scharnhorststraße 20
Leipzig
DDR-7030

Dr. H. NEUMANN
Sektion Physik
Karl-Marx-Universität
Linnéstraße 5
Leipzig
DDR-7010

Dr. M. S. OMAR
School of Physics
University of Bath
Claverton Down
Bath BA2 7AY
United Kingdom

Cryst. Res. Technol.	23	1988	1	107–109

H.-J. Höbler, A. Zahn

Sektion Chemie, WB Kristallographie der Karl-Marx-Universität Leipzig

Crystallographic Investigations of Glaucochroite in an Fe–Mn Slag

Dedicated to Professor Hermann Neels on the occasion of his 75th birthday

1. Introduction

Calcium silicates are commonly occuring compounds in steel slags. In industrial slags available in GDR the variety of minerals is practically restricted to dicalciumsilicate (β-C_2S, γ-C_2S), tricalciumsilicate (C_3S), monticellite (CMS), merwinite (C_3MS_2) and forsterite (M_2S). Up to now glaucochroite CaMn[SiO_4] in these slags was not observed. In accordance with Trojer this mineral forms in steel slags as a solid solution with kirschsteinite (CfS), monticellite, tephroite (Mn_2S) or γ-C_2S. Only naturally originated glaucochroite exists in the pure form. It is the purpose of this work to report on the occurrence of a relative pure glaucochroite in a slag produced in GDR and its crystallographic properties. The slag forms in the arc furnace on the production of Fe–Mn-alloys. Mn ores serving as raw material are reduced by a Si–Mn alloy. The Fe–Mn alloy and the slag are seperately taped in chills.

2. Experimental and results

The greenish-grey coloured slag macroscopically shows some small longish pores. Metallic inclusions are also observable. Microscopically 3 phases can be recognized (see Fig. 1): The main component consists of longish crystals up to 5 mm in length appearing colourless and anisotropically in thin section. In reflected light these crystals only can be separated by means of a less reflecting phase between them. Further, lightly reflecting dendritic formations are discernible. In thin section these dendrites are isotropic and green diaphanous. Chemical compositions determined by ESMA measurements are listed in Table 1.

Fig. 1. Glaucochroite crystals and manganosite dendrites in a Fe–Mn slag, thin section, crossed nicols

Table 1
Phase compositions measured by ESMA

glaucochroite	$Ca_{0.96}Mn_{0.98}Mg_{0.06}SiO_4$ $Ca_{0.99}Mn_{0.99}Mg_{0.04}SiO_4$
manganosite	MnO
intermediate phase	enrichment of Na_2O, K_2O, BaO

Table 2
X-ray characteristic of Fe–Mn slag

Observed reflections		glaucochroite (nat.)			manganosite	
d	I/I_0	*d*	I/I_0	*hkl*	*d*	I/I_0
5.60		5.54	30	020		
4.24	24	4.23	40	021		
3.93	9	3.91	10	101		
3.71	41	3.69	60	111, 120		
3.26	3	3.25	20	002		
2.97	84	2.96	60	130		
2.82	17	2.81	20	022		
2.80	5			040		
2.71	66	2.69	80	131		
2.64	100	2.63	80	112		
2.567	29	2.56	20	041	2.56	62
2.445	10	2.44	30	122		
2.426	10					
2.259	5	2.25	20	211, 220		
2.219	22				2.223	100
		2.19	20			
2.038	4	2.03	20	150		
1.959	3	1.95	20	113, 231		
1.854	78	1.85	100	222, 240		
1.792	7	1.79	20	061		
1.781	6	1.77	20			
1.748	9	1.75	20			
1.726	7	1.72	20			
		1.71	20			
1.618	11	1.62	40			
		1.61	40			
1.572	4	1.58	20		1.571	58
1.520	11	1.52	30			
1.455	3					
1.430	2	1.43	20			
1.421	3	1.42	30			
1.339	6				1.340	21
1.284	5				1.283	13
1.225	3	1.23	20			
1.195	0.6	1.19	20			
		1.14	20			
1.130	8	1.13	50			

The composition of main phase is very close to the orthorhombic mineral glaucochroite $CaMn[SiO_4]$ crystallizing in an olivine like structure, that of dendrites to manganosite MnO. In the intermediate phase Na_2O, K_2O, and BaO are enriched. To conform this result optical and X-ray characteristics were accomplished. X-ray measurements were performed at powder diffractometer HZG 4, using CuKα radiation and Si as internal standard. The step width of the detector was $2\theta = 0.05°$. The positions and relative intensities of reflection are listed in Table 2. They correspond to a mixture of glaucochroite (powder diffraction file 14–376) with some manganosite (powder diffraction file 7–230). Lattice parameters of glaucochroite were ascertained using four suitable triplets of

Table 3
Optical properties and lattice parameters of glaucochroite and γ-$Ca_2[SiO_4]$

	glaucochroite (observed)	glaucochroite (file 14-376)	γ-$Ca_2[SiO_4]$ TROJER)
n_β	1.713 ± 0.003	1.723	1.645
Δ	0.055 ± 0.005	0.051	0.012
$2V$	$-67 \pm 1°$	$-61°$	$+60°$
a_0	4.94 ± 0.01 Å	4.92 Å	5.06 Å
b_0	11.18 ± 0.02 Å	11.14 Å	11.28 Å
c_0	6.52 ± 0.01 Å	6.50 Å	6.78 Å

reflections for calculation and finding out the arithmetic mean. For optical characterization of glaucochroite thin sections of the Fe–Mn slag were examined under the polarizing microscope combined with an universal stage. Index of refraction n_β, birefringence Δ, and the angle between the optical axes $2V$ were determined. These values and the lattice parameters are listed in Table 3 (left column).

3. Discussion

For comparison of these results the values of natural glaucochroite (powder diffraction file 14-376) are cited (middle column of Tab. 3). It is noticed that the index of refraction n_β and the angle between optical axes $2V$ clearly deflect from pure glaucochroite. The observed lattice parameters are slightly enlarged. These differences can not be explained by a solid solution formation in the system $CaMn[SiO_4]$–$CaFe \cdot [SiO_4]$–$CaMg[SiO_4]$ cited by TRÖGER, because the lattice parameters in this system are always smaller than those of pure glaucochroite. Similarly, solid solutions with the olivine type minerals forsterite, fayalithe and tephroite would result in a reduction of elementary cell. Its enlargement can be explained by a solid solution formation with γ-$Ca_2[SiO_4]$, also crystallizing in the olivine type. For comparison the crystallographic data of this compound are listed in the right column of Table 3. The observed differences in optical properties can also be attributed to this solid solution formation, resulting in a diminution of indices of refraction and, due to the opposite optical characters of both minerals, in an enlargement of $2V$. However, quantitative statements on the content of γ-C_2S cannot be given, because the dependence of the determined values on composition is not investigated up to now.

The authors are grateful to Dr. P. Lange (Jena) for accomplishing ESMA investigations to Mr. G. Kommichau for carrying out X-ray measurements and to Dr. P. Schreiter for helpful discussions.

References

TRÖGER, W. E.: Optische Bestimmung der gesteinsbildenden Minerale, Teil 1, Bestimmungstabellen, Stuttgart 1982

TROJER, F.: Die oxydischen Kristallphasen der anorganischen Industrieprodukte, Stuttgart 1963

(Received July 10, 1987)

Authors' addresses:

Dr. H.-J. HÖBLER, Dr. A. ZAHN
Sektion Chemie, Wissenschaftsbereich Kristallographie, Karl-Marx-Universität
Scharnhorststraße 20
Leipzig
DDR-7030

Cryst. Res. Technol.	23	1988	1	111–115

W. Schmitz, P. Schreiter

Wissenschaftsbereich Kristallographie der Sektion Chemie der Karl-Marx-Universität Leipzig

X-ray Determination of Solid Solution Composition in the Melilite System Åkermanite (AK) — Gehlenite (Ge) — Soda Melilite (Sm)

Dedicated to Professor Germann Neels on the occasion of his 75th birthday

Expressions are given and discussed which allow a determination of the composition of solid solutions in the ternary system Ak–Ge–Sm. Resulting from X-ray measurements of the tetragonal melilite constants a_0 and c_0, the composition can be calculated within the limits of $\pm 3\%$ if the lattice constants are determined with an accuracy of ± 0.001 Å.

Es werden Lösungsgleichungen für die Bestimmung der Mischkristallzusammensetzung im ternären System Ak–Ge–Nm abgeleitet und diskutiert. Aus der röntgenographischen Messung der Gitterkonstanten a_0 and c_0 der tetragonalen Melilithe ist es möglich, die Zusammensetzung in den Grenzen von $\pm 3\%$ zu berechnen, wenn die Konstanten mit einer Genauigkeit von ± 0.001 Å bestimmt werden.

1. Introduction

Melilites belong to a tetragonal mineral group ($P\bar{4}2_1m$) being very complex in composition. Their general formula has been determined to

$X_2YZ_2O_7$ (Warren) .

They form solid solutions in a very broad range and following the relevant investigations, it is possible to assign to

X = Ca, Na, Sr; K (partly)

Y = Mg, Al, Fe^{2+}, Fe^{3+}, Mn, Ti, Zn, Ga

Z = Si, Al, Ga (Smith; Christie; Nurse, Midgley; Raaz, Schichtl)

The melilite structure (Warren; Raaz, Schichtl; Louisnathan) mainly consist of simple tetrahedra (YO_4) and double tetrahedra (Z_2O_7) arranged to a tetragonal framework whereby the X-atoms are asymmetrically surrounded by eight O-atoms.

Apart from their occurrence as minerals in the earth's crust, melilites play an important rôle as phase components of industrial by-products as slags.

From this point of view we have to consider the solid solutions of the end-components of:

Åkermanite (Ak) $Ca_2MgSi_2O_7$
Gehlenite (Ge) $Ca_2Al(AlSi)O_7$
Soda melilite (Sm) $(CaNa)AlSi_2O_7$.

In order to produce ceramic materials on the base of melilite, it is necessary to know the chemical composition of melilite solid solutions to be able to influence the formation of melilite ceramic under technical conditions.

2. Experimental

Melilite specimens were synthesized from the starting materials CaO or $CaCO_3$, MgO or $MgCO_3$, Na_2CO_3, Al_2O_3, SiO_2 (all chemicals p.a.). The oxides quantitatively turn to melilite over solid state reactions at 1100 °C during 220 h. The preparation in platinium cruibles with the aid of solid state reactions was choosen to avoid occurrence of a molten phase which would be difficult to analyse. The end components Ak and Ge and also their solid solutions could be prepared in this way. All samples were checked for pureness by optical microscopy and X-ray methods. A weight loss could not be observed. In the case of Sm, we were not able to prepare solid solutions of definite composition.

X-ray measurements were carried out by the powder diffractometer HZG4 (Freiberg, GDR). The lattice constants of the melilite specimens were determined from $hk0$ and $00l$ reflections using Si as internal standard (reproducibility ± 0.001 Å).

3. Results and discussion

Table 1 shows the lattice constants determined for Ak and Ge and their solid solutions of defined compositions. Figures 1, 2 give the behaviour of a_0, c_0 in the system Ak–Ge showing marked deviations from the Vegard line.

Table 1
Lattice constants (Å) in the system åkermanite – gehlenite

Composition	a_0	c_0
Ak	7.832	5.006
$Ak_{80}Ge_{20}$	7.798	5.023
$Ak_{60}Ge_{40}$	7.764	5.035
$Ak_{40}Ge_{60}$	7.732	5.049
$Ak_{20}Ge_{80}$	7.708	5.058
Ge	7.686	5.065

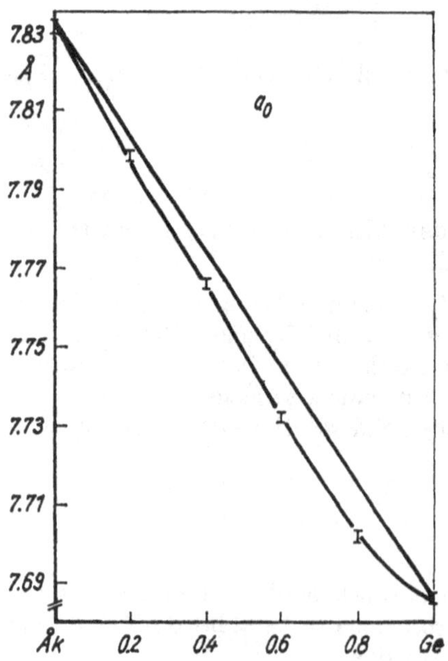

Fig. 1. Behaviour of a_0

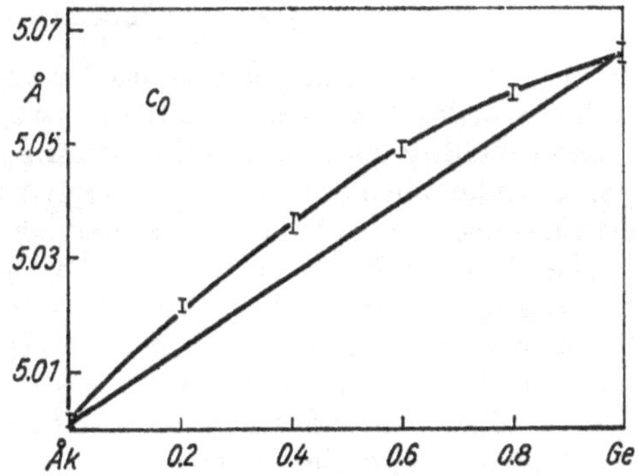

Fig. 2. Behaviour of c_0

The opposite deviations from the Vegard line concerning a_0 and c_0 can be explained by the parameters of the melilite structure. The results of X-ray structure determination have shown an ordered distribution of the Al^{3+} and Si^{4+} in the Z_2O_7 double tetrahedra for Ge (LOUISNATHAN 1971). This order cannot be maintained in solid solutions. The substitution of Mg^{2+} for Al^{3+} in the YO_4 tetrahedra is connected with a substitution of Si^{4+} for Al^{3+} in the Z_2O_7 double tetrahedra. As a result of different ionic radii in different sites of the structure, an opposite slope of the curves has been experimentally obtained for the lattice constants in the Ak–Ge system. The deviations from the Vegard line result from exchange process concerning both kinds of tetrahedra in each case of solid solution formation in the ternary system Ak–Ge–Sm. Therefore, similar deviations from the Vegard line can be predicted in the non-measured system Ak–Sm.

The knowledge of the lattice constants slope is necessary of both binary systems in order to determine the composition of any solid solution in the ternary system by means of lattice constants measurements.

4. Determination of the solid solutions composition

4.1. Binary system akermanite — gehlenite

These results originate in measured lattice constants (cf. Tab. 1). The behaviour of the constants can be expressed by quadratic functions:

$$y_a = 0.044x^2 - 0.190x + 7.832\,, \qquad (1)$$

$$y_c = -0.031x^2 + 0.090x + 5.006\,. \qquad (2)$$

They give the lattice constants if the molecular fraction x_{Ge} is known.

One of the two solutions of the quadratic equations (3, 4), derived from 1, 2, give the solid solution composition (x_{Ge}) in the Ak–Ge system:

$$x^2 - 4.32x + (7.832 - y_a)/0.044 = 0 \qquad (3)$$

$$x^2 - 2.90x - (5.006 - y_c)/0.031 = 0 \qquad (4)$$

Best degree of accuracy is reached ($\pm 2\%$) when the calculation is based on a_0 constant because of its larger lattice constants difference between the end components (cf. Tab. 1).

4.2. Binary system åkermanite — soda melilite

In order to determine solid solution composition in a ternary system, it is necessary to know at least lattice constants of two binary sets.

Soda melilite containing solid solutions only can be prepared to $Ak_{43}Sm_{57}$ or $Ge_{30} \cdot Sm_{70}$ under normal conditions (NEUVONEN). They tend more to coexist with other crystalline or amorphous phases than Ak–Ge.

Therefore we had to construct the Vegard lines from measured constants of Ak and those for soda melilite according to SCHICHTL and RAAZ and to LOUISNATHAN (1970) who synthesized the mineral under high pressure.

Since heterovalent substitution leads to deviations from the Vegard line in the system of Ak–Ge and a similar situation has to consider for Ak–Sm, we think it is justified to assign the deviations as they were found for Ak–Ge. Adopting this approach, the following expressions were obtained for the Ak–Sm system

$$y_a = 0.043x^2 - 0.244x + 7.832 \quad (5)$$

$$y_c = -0.028x^2 + 0.070x + 5.006 \quad (6)$$

and

$$x^2 - 5.67x + (7.832 - y_a)/0.043 = 0 \quad (7)$$

$$x^2 - 2.50x - (5.006 - y_c)/0.028 = 0 \quad (8)$$

(7) and (8) give x_{Sm}.

4.3. Ternary system åkermannite — gehlenite — soda melitite

An example of the calculation of the solid solution composition in the ternary system Ak–Ge–Sm is demonstrated by Figure 3.

Using cartesian coordinates (use of triangle coordinates is also possible), the ratio x_{Sm}/x_{Ak} and x_{Ge}/x_{Ak} is entered in the diagram. The plot of two straight lines (one for a_0 and one for c_0) has to be made resulting from the intersection with the axis of coordinates (Fig. 3). These intersection points are calculated from the relations shown in (3) giving a_2 and (7) giving b_2 for a_0 and (4) giving a_1 and (8) giving b_1 for c_0. They indicate the molecular ratio x_{Ge}/x_{Ak} (3, 4) and x_{Sm}/x_{Ak} (7, 8) for the binary systems.

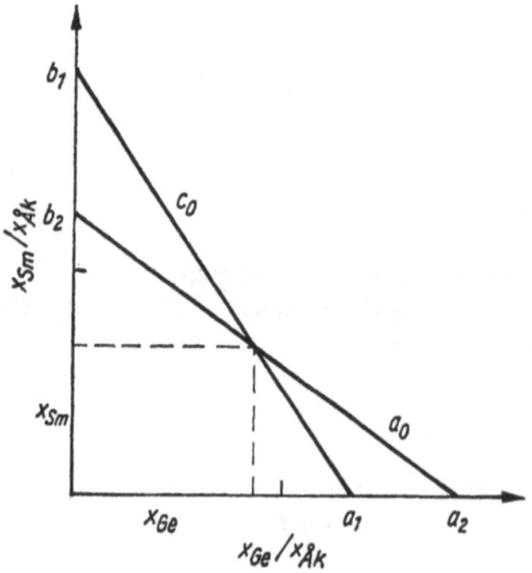

Fig. 3. Construction of the straight lines for a_0 and c_0 resulting from the relations (3, 4, 7, 8). The intersection point (marked by dotted lines) provides the solid solution composition in the ternary system

After all, the intersection point of the two lines provides the composition of the solid solution in the ternary system Ak–Ge–Sm by

$$x_{\mathrm{Sm}} + x_{\mathrm{Ge}} + x_{\mathrm{Ak}} = 1 .$$

x_{Sm} and x_{Ge} can be read off from the diagram or can be calculated by mathematical methods.

According to the relation of a straight line

$$x/a + y/b = 1 ,$$

it is possible to calculate the intersection point of the two lines by:

$$x_{\mathrm{Ge}} = \begin{vmatrix} 1 & 1/b_1 \\ 1 & 1/b_2 \end{vmatrix} : \begin{vmatrix} 1/a_1 & 1/b_1 \\ 1/a_2 & 1/b_2 \end{vmatrix},$$

$$x_{\mathrm{Sm}} = \begin{vmatrix} 1/a_1 & 1 \\ 1/a_2 & 1 \end{vmatrix} : \begin{vmatrix} 1/a_1 & 1/b_1 \\ 1/a_2 & 1/b_2 \end{vmatrix}.$$

The composition of a melitite sample in the ternary system Ak–Ge–Sm can be determined in this way with a reproducibility of $\pm$ 3% if the lattice constants are measured with an accuracy of $\pm$ 0.001 Å.

References

Christie, O. H. J.: Norsk geologisk tidskrift **42**, 31 (1962)
Louisnathan, S. J.: Z. Krist. **131** (1970)
Louisnathan, S. J.: Can. Mineral. **10**, 822 (1971)
Neuvonen, K. J.: Bull. Commiss. geol. Finlands **158**, 16 (1952)
Nurse, A., Midgley, W.: J. Iron Steel Inst. London **174** (1953)
Raaz, F., Schichtl, H.: Anzeiger der Österreich. Akad. der Wissensch. **15**, 361 (1968)
Smith, J. U.: Amer. Min. **38**, 643 (1953)
Warren, B. J.: Z. Krist. **74**, 131 (1930)

(Received July 15, 1987

Authors' address:

Dr. W. Schmitz, Doz. Dr. P. Schreiter
Wissenschaftsbereich Kristallographie, Sektion Chemie der Karl-Marx-Universität
Scharnhorststr. 20
Leipzig
DDR-7030

Cryst. Res. Technol.	23	1988	1	117–126

E. KURTH, A. REIF, V. GOTTSCHALCH, J. FINSTER, E. BUTTER

Karl-Marx-Universität Leipzig, Sektion Chemie, Leipzig, DDR

Chemical Etching and Polishing of InP

Dedicated to Professor NEELS on the occasion of his 75th birthday

This paper describes possibilities of several chemical preparations for the selective cleaning of InP surfaces. The investigations of the surface states after the chemical treatment were carried out by means of XPS measurements. A pre-etching with $(NH_4)_2S_2O_8:H_2SO_4:H_2O$ and a polishing with 1% bromine in methanol produce optically smooth (100)- and (111) P surfaces free of oxides.

In dieser Arbeit sind verschiedene Möglichkeiten der selektiven Reinigung von InP-Oberflächen aufgezeigt. Die Untersuchungen der Oberflächenzustände nach der chemischen Behandlung wurden mittels XPS-Messungen durchgeführt. Mit einem Vorätzer, bestehend aus $(NH_4)_2S_2O_8:H_2SO_4:H_2O$ und einem Poliermittel aus $Br_2:CH_3OH$ (1%) können glatte und oxidfreie (100)- und (111) P-Oberflächen erhalten werden.

Introduction

InP has many potential applications in semiconductor devices, such as microwave devices, high-frequency FETs and semiconductor diode lasers. Efficiently and performance of such devices depend critically on the surface preparation (SING et al.). For instance, plane and smooth surfaces free of any damage and chemical impurities are required for LPE growth of $In_{1-x}Ga_xAs_yP_{1-y}P_y$/InP double heterostructure. Epitaxial growth on oxidized surfaces or on surfaces contaminated with impurities can result in holes or other forms of irregularities within epitaxial layers (TRAPP, ERMANIS), because surface oxides hinder the melt wetting of the substrate (KÖNIG et al.; SCHWARTZ et al.).

The most commonly used method to remove the ground polish damage is the fine polish with Br_2 in CH_3OH. However, the chemical reactivity of bromine is not sufficient to resolve all impurities from the InP surface. Thus, etch hillocks are often formed. Many authors investigated a pre-etching to remove the insoluble impurities before they used Br_2 methanol: KOH (FIEDLER, SCHLACHETZKI), HNO_3 (KENZO et al.), $H_2SO_4:H_2O_2:H_2O$ (NAKAJIMA et al.; NISHITANI, KOTANI). The application of both systems $CH_3OH:Br_2$ and $H_2SO_4:H_2O_2:H_2O$ is connected with difficulties, because those systems are reactive mixtures the compositions and etch effect of which depend on their preparation conditions as temperature and time after beginning of preparation. Moreover, we should expect much more complicated chemical surface reactions due to the high ionicity (40% (PHILLIPS)) of the semiconductor compound InP.

This paper tries to show possibilities of several chemical preparations for the selective cleaning of InP surfaces of the important orientations (100), (111) In, (111) P. The investigations of the surfaces were carried out by means of XPS measurements and optical interference contrast microscopy.

Experimental

The surfaces of tin doped InP samples were polished and of the orientations (100), (111) In, and (111) P (free carrier concentration roughly 1×10^{10} cm^{-3}). The samples were treated in a PTFE-vessel as listed in Table 1. The change of etchants and cleaning solutions were carried out by a quick displacement of the solutions without air contact of the samples. The exposure to air of freshly etched and rinsed InP surfaces was restricted to a maximum of 1 minute until they placed in the vacuum chamber of the spectrometer Vacuum Generators ESCA 3. After loading, the vacuum chamber was quickly evacuated to a pressure of 10^{-8} torr. All binding energies were corrected to a carbon 1s binding energy of 285.0 eV to compensate the effects of sample charging. The precision of binding energies was determined to $\leqq 0.15$ eV. The relative mistake of the measured peak intensities was calculated to 6%. All intensities were corrected by means of tabulated values (Bindungsenergie-Katalog). In addition, the binding energy data of the surface compounds assumed were compared with published data (WAGER et al.; CLARK et al.; KAZMERSKI et al.; WILMSEN, KEE; NELSON et al.; BERTRAND; WALDROP et al.).

Table 1
Treatments of InP surfaces and the XPS investigations

Treatment	conditions	spectrum
1. Rinsing		
trichlorethylen		
aceton		
methanol	RT	A
2. Pre-etching		
HNO_3 (65%):H_2O	RT/1 min	B
KOH (30%):H_2O	40 °C/4 min	C
$(NH_4)_2S_2O_8$:H_2SO_4:H_2O		
15:73:15	RT/1 min	D/G/H
7:73:25	RT/1 min	E
15:73:25	RT/1 min	F
3. Rinsing		
methanol		
4. Polishing		
Br_2:CH_3OH (1%)	RT/2 min	
5. Rinsing		
methanol	RT/2 min	K/L/M
6. Oxid-free etching		
HCl:CH_3OH (1:10)	RT/10 s	
7. Rinsing		
H_2O	RT/5 min	R
CH_3OH	RT/1 min	Q

Properties of the system H_2SO_4:H_2O_2:H_2O

Peroxomono- and peroxodisulfuric acid (H_2SO_5 = Y; $H_2S_2O_8$ = X) are formed as secondary reaction products after mixing the system H_2SO_4 (Z):H_2O_2:H_2O. The concentrations of all included components depend on the time after mixing, temperature, starting concentrations, and impurities. PALME investigated the progress of hydrolysis reactions of $H_2S_2O_8$ (H1; H2).

$$H_2S_2O_8 + H_2O \underset{k_2}{\overset{k_1}{\rightleftharpoons}} H_2SO_5 + H_2SO_4 \quad (H1)$$

$$H_2SO_5 + H_2O \underset{k_4}{\overset{k_3}{\rightleftharpoons}} H_2SO_4 + H_2O_2 \quad (H2)$$

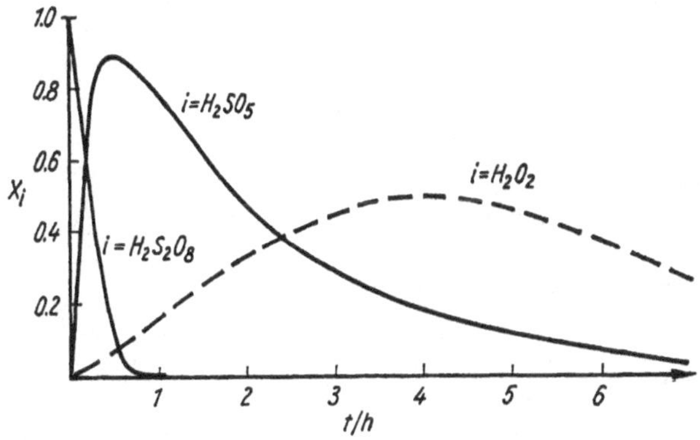

Fig. 1. Variation of composition from a mixture $(NH_4)_2S_2O_8:H_2SO_4:H_2O$ after preparation begin

He determined the ratios of velocities constants k_1/k_3 and k_4/k_3 for H_2SO_4 concentration > 2 M. Additionally, reactions between the components listed, forming O_2 and O_3, have to be taken into consideration. In relation to the oxidation potential, H_2SO_5 is much more reactive than H_2O_2. $H_2S_2O_8$ reacts very slowly as oxidant. Figure 1 presents the principle composition development of a mixture of $(NH_4)_2S_2O_8:H_2SO_4:H_2O$. There is observed a concentration maximum of each peroxide compound. The hydrolysis reaction (H1) slows down proportional to the concentration of sulfuric acid. The preparations of mixtures D to F must be performed carefully draining the heat developed by the dilution of sulfuric acid. The oxidation power of H_2SO_5 is sufficient to oxidize $C_xH_yO_z$ compounds to CO_2 and H_2O. An additional advantage is that there do not exist any insoluble peroxodi- and peroxomonosulfates in the sulfuric acidic media.

Results and discussion

We have freshly prepared mixtures of several compositions. Then we have etched InP samples after rinsing with trichlorethylen, aceton and methanol, following by rinse with distilled water (conductivity $\sim 1\ \mu S$). The measured XPS spectra are compared in Figure 2, the spectra of the samples pretreated with KOH and HNO_3 (Tab. 1) too.

The binding energy of P (2p) in InP (129.5 ± 0.1 eV) is identical for all samples etched. The binding energy of P (2p) is 4.5 eV higher in structure fragments as $[PO_4^{3-}]$. On the other hand, the P (2p) binding energies of HNO_3 etched sample B and of the sample E, F are only 3.5 to 4.2 eV higher. We suppose that this lower concentrated acids favour the formation of acidic phosphates $[HPO_4^{2-}]$. The reaction of $KOH:H_2O$ with InP seems to be complex like the reaction with elemental phosphorus (Remy). The shape of the spectrum C points to the existence of a thin layer of phosphorus compounds at various oxidation stages. Contaminations of Si-compounds could not be dissolved entirely. The treatment with concentrated nitric acid gives heavily oxidized surfaces which is in agreement with the results of Bertrand. The a $(NH_4)_2S_2O_8$:b H_2SO_4:c H_2O etch leaves a thin oxide layer if the composition ratio c:a is greater than one, or if the etch mixture is older than 3 h.

That means, higher concentrations of H_2O_2 and H_2O in relation to H_2SO_4 accelerate the oxidation of InP surfaces. Contrary to that higher concentrations of H_2SO_5 in

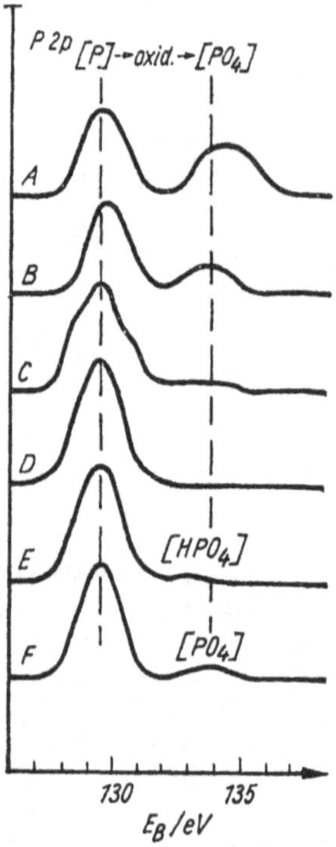

Fig. 2. XPS-spectrum of P 2p-level from (100) InP surfaces after different pre-etching (etching parameters see Table 1)

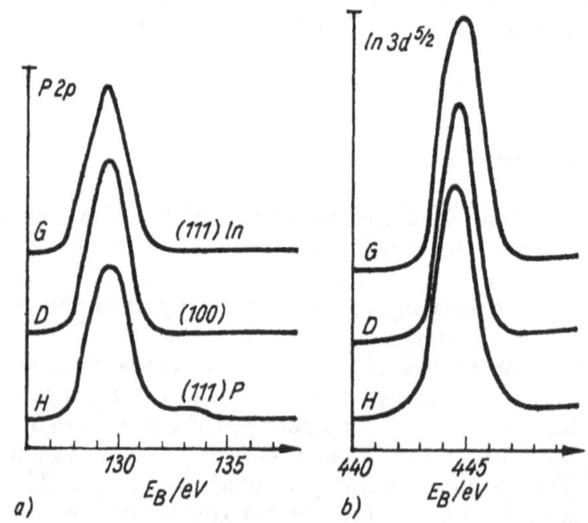

Fig. 3. Spectra of binding energy of P 2p (a) and In $3d^{5/2}$ (b) from samples with different orientation (G-(111) In, D-(100), H-(111) P)

relation to H_2O_2 and H_2O effect more selective dissolution of oxides and other surface impurities without attack of InP. The etch rate investigations made by NISHITANI seem to confirm this assertion. The etch rate increases proportional to the H_2O_2/H_2O ratio. This can be due to the reaction of H_2O_2 in diluted sulfuric acidic media followed by electrophile attack onto the phosphide ion:

$$H_2O_2 + H_2SO_4 \rightleftharpoons H_3O_2^+ + HSO^-$$

$$InP + H_3O_2 \rightarrow [InPO]_{(m)} + H_3O^+$$

H_2SO_5, a monovalent acid, does not react in this way. Whereas the mixture D removes all oxides from (100) InP selectively, we have found structure fragments as P–O compounds on (111) P surfaces and [InOH] and [SO_4] on (111) In faces (Fig. 3). Both results are either due to the oxidation by atmospheric humidity or adsorbed ions (for instance H^+-ions) obtained by the rinse (dilution and reaction) with H_2O after etching. The application of rinse with methanol (conductivity $\sim 1\,\mu S$) for 1 h immediately after etching with mixture D leads to better results. The measured spectra of the orientations (111) In and (111) P show the same shape as spectrum D for (100) InP. The binding energy found for chemisorbed oxygen (E_B (O is) $\sim$ 532.5 eV) may be belong to water adsorbed before loading in the vacuum chamber. The carbon contamination seems not to be higher than the contamination of surfaces sputter-cleaned with Ar^+ (−80 Å). The selective removing of the hardly soluble oxides as $C_xH_yO_z$-compounds and other impurities with mixture D followed by methanol rinse allows to produce by means of Br_2 (1 vol%):CH_3OH polish final optical smooth (100) and (111) P surfaces. The etching rate of Br_2:CH_3OH is large enough to remove the damaged surface layer. After the Br_2:CH_3OH touch with (100)-InP the surfaces have adsorbed more than one monolayer of bromine. The other orientations (111) In, (111) P are able to hold only few bromine for a short time. The bond stages of adsorbed bromine change with time. Figure 4 shows the time dependence of the Br (3d) signal intensity calibrated in relation to the signal intensity of In (4d) under room air conditions (60% relative atmospheric humidity; 20 °C; atmosphere did not contain any bromine). The volatilization of some of adsorbed bromine (bond stage I) in the vacuum chamber is probably due to the bond type between halogen molecules. Such

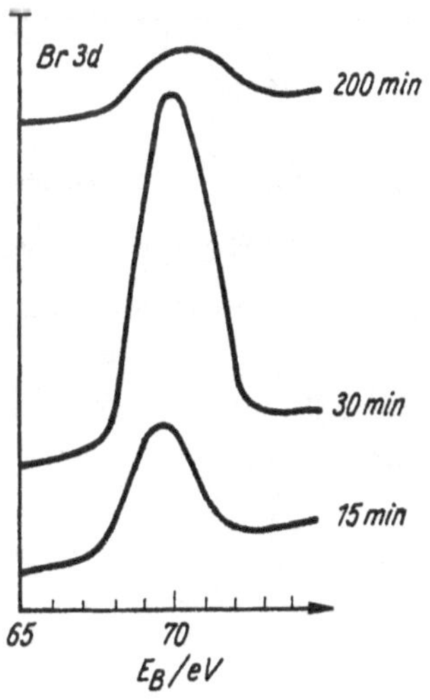

Fig. 4

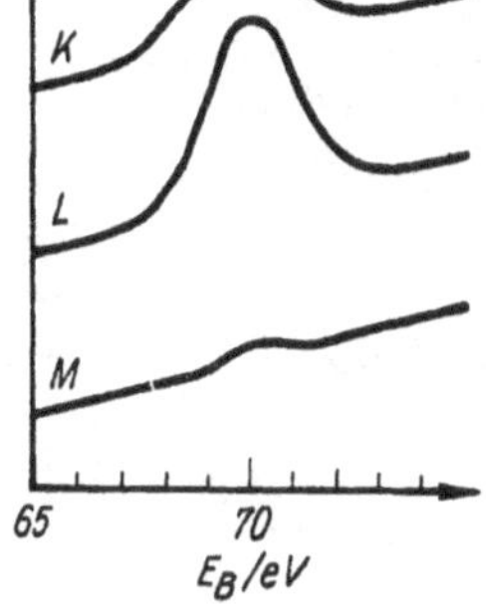

Fig. 5

Fig. 4. Change of the Br (3d)-signal of etched and rinsed sample, which is placing different time under room air conditions

Fig. 5. Spectra of binding energy of Br (3d) from different oriented samples after Br_2:CH_3OH-polishing (K-(111) In, L-(100), M-(111) P)

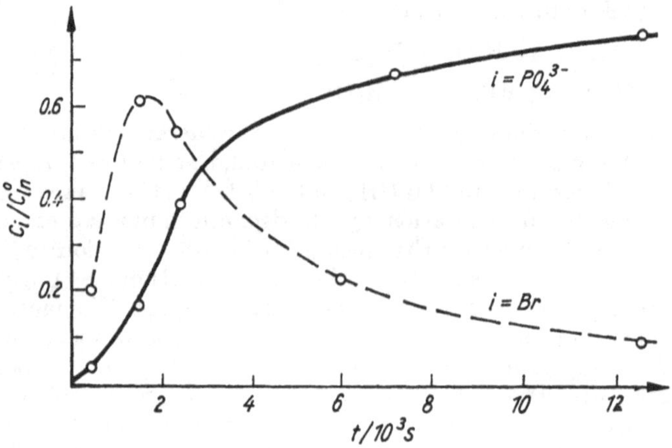

Fig. 6. Growth of the $(PO_4)^{3-}$ concentration in dependence on the Br_2 concentration

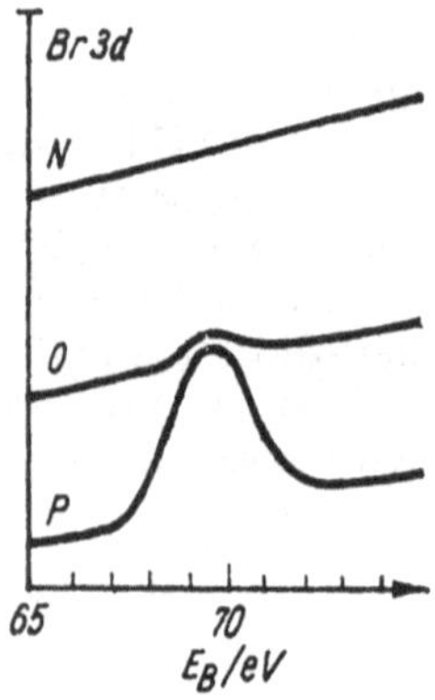

Fig. 7

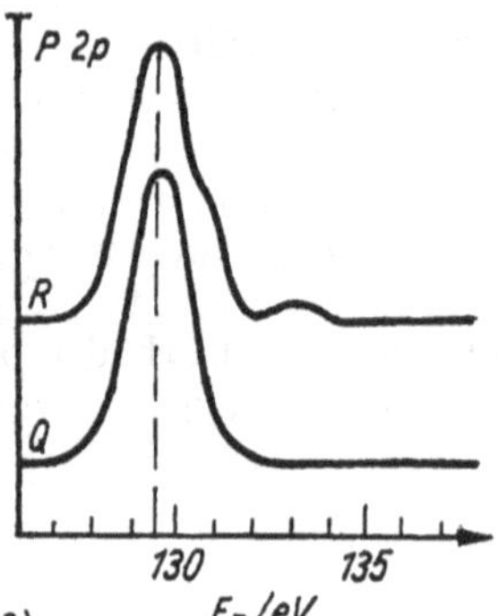

8)

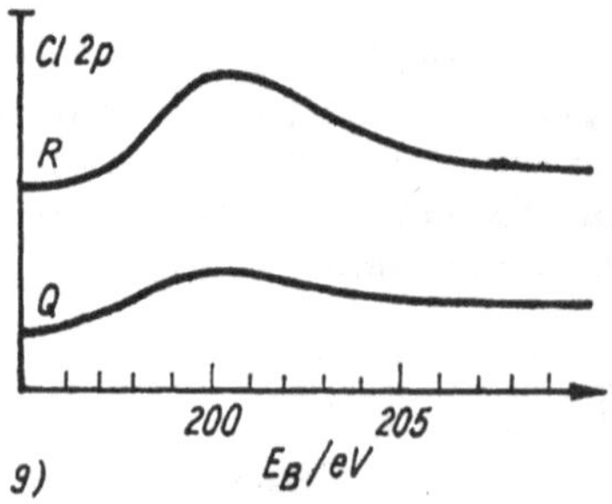

9)

Fig. 7. Decreasing of the Br_2 concentration with the rinsing time (CH_3OH), N – 100 min O – 30 min P – 15 min

Fig. 8. Spectra of binding energy (P 2p) after etching with $HCl:CH_3OH$ and rinsing with H_2O (R) and CH_3OH (Q)

Fig. 9. Spectra of binding energy (Cl 2p) after etching with HCl: CH_3OH and rinsing with H_2O (R) and CH_3OH (Q)

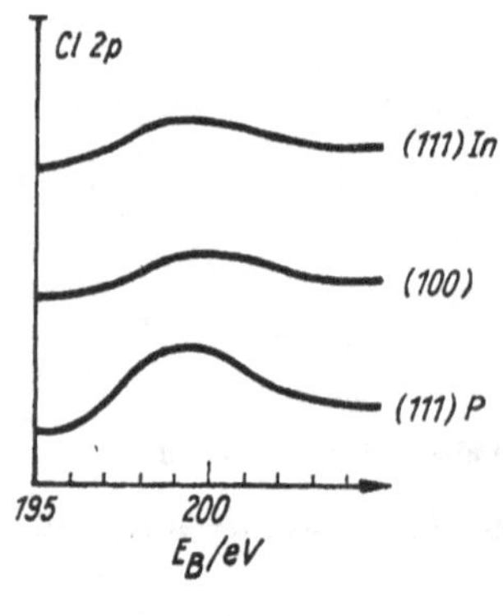

Fig. 10. Orientation binding energy (Cl 2p) after etching with $HCl:CH_3OH$ and rinsing with H_2O (R) and CH_3OH (Q)

Fig. 10. Orientation dependence of chlorine adsorption

bromine associates crack and bromine will be adsorbed directly at the surface. The bromine adsorption on the (100) plane is more stable than on (111) InP (Fig. 5). However, we have observed only a few traces of bromine on (111) P. The relatively high value of Br (3d) binding energy (70 eV) hints Br–In charge-transfer complex bonds. The bromine stabilization on (100) can be to explained with similar In–Br–In bridge bonds as in the case of In_2Br_6. It is known that indium bromide compounds undergo hydrolysis reactions with H_2O. In this way bromine adsorbates are decomposed leading to a very fast catalytic surface oxidation as shown in Figure 6. The surface oxidation can be retarded by rinsing and keeping the samples in methanol. Then bromine desorbs from the surface outlined by Figure 7 due to a first order reaction. The bromine desorption is finished after 90 minutes. However, traces of water (formed by secondary reactions of Br_2 with H_2O adsorbed from atmosphere or with CH_3OH itself) in methanol effect the formation of measurable amounts of surface oxides. The conscious additions of water (< 8 vol%) accelerate the oxidation. Nelson and other workers had used a short time etching with the solution HCl: 10 H_2O followed by a H_2O rinse for selective removing of surface oxide traces from (100) InP. The spectrum R in Figure 8 shows a peak shoulder at 130.5 eV and a second peak at 133.3 eV. In the following these results are compared with those of short time etching with HCl:CH_3OH. The latter provides (100) InP surfaces free of oxides (spectrum Q), respectively.

Beside the shifting of P (2p) binding energy to higher values the detected chlorine contamination is better suitable to estimate the adsorption grade of hydrogenchloride and therefore the H^+-ion concentration too. Figure 9 shows the binding energy intensities of the Cl (2p) level. The orientation dependence of chlorine adsorption is recorded in Figure 10. The Cl (2p) binding energy of about 200 eV does not represent any of chloride compounds, but it is probably attributed to a non-ionic stage of chlorine in adsorbed HCl like this type $[P^-–H^+–Cl^-]$. This relationship is supported by the orientation dependence of chlorine concentration. The highest chlorine concentration was found on (111) P, the lowest on (111) In planes. In such a manner these adsorbed compounds will be stabilized much better by H_2O to complexes than by CH_3OH as depicted in Figure 9.

The presented XPS results about surface chemical reactions of InP indicate clearly that the last chemical treatment including the immediately followed rinse define the surface chemical composition. Type and quantity of adsorbed impurities depend also on the orientations. Several simple impurities seem to own a deciding influence on the stability of surfaces and especially on the velocity of following chemical surface reactions.

Stability of oxide-free etched InP surfaces at room conditions

It is generally interesting to know which treatment provides a minimum of oxidation velocity. Chye et al. studied the initial oxidation of cleaned (110) InP surfaces using photoemission. They have observed that this surface was extremely resistant to oxidation. However, if it finally reacted with oxygen. there was no intermediate chemisorption stage as observed in the initial oxidation of GaAs.

Wager, Wilmsen investigated growth rates and chemical composition of thermally grown (temperature range 340 to 450 °C) oxides on (100) InP in dry oxygen. They found by means of XPS measurements In_2O_3 rich surfaces. The growth kinetic investigations were started after the formation of oxide layers with 12 to 20 Å thickness. Other authors ascertained the chemical changes by thermally oxidizing (100 to 730 °C) by XPS, UPS, and SIMS (Kazmerski et al.; Schwartz et al.).

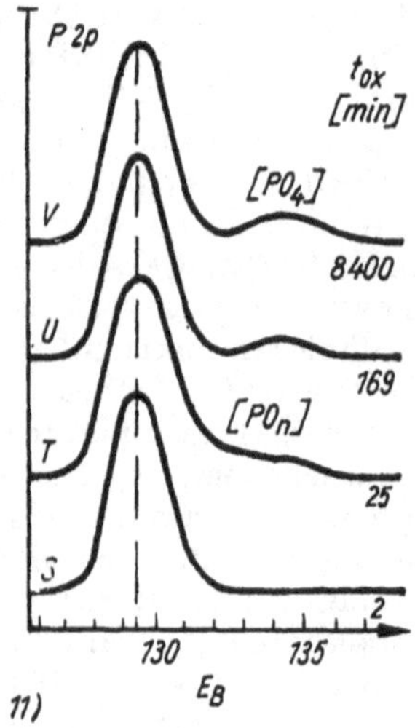

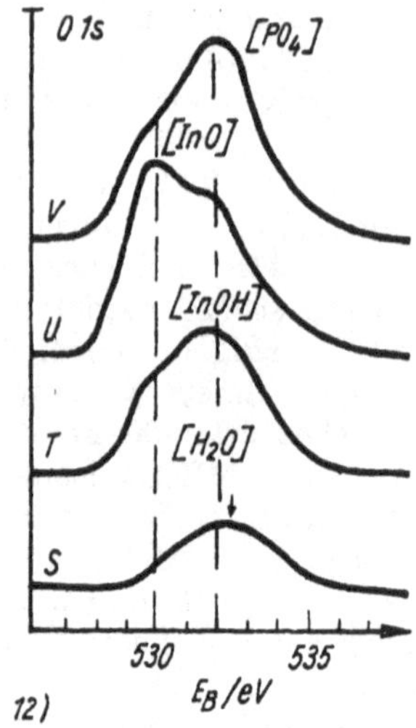

Fig. 11. Spectra of binding energy (P 2p) in dependence on oxidation/time
Fig. 12. Changing of binding energy (O is) on the time of atmospheric oxidation

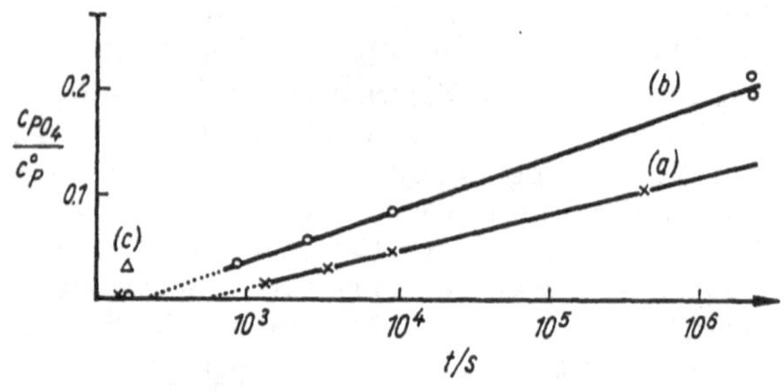

Fig. 13. Functionally dependence of oxide growth after different etching treatments

Up to now, there are no papers on oxidation mechanism at room temperature and its dependence on residual contaminations left by defined wet treatments.

Therefore the surface compositions were investigated after selective removing of oxides by the following treatments:

(a) mixture D, 30 s methanol rinse
(b) mixture D, 30 s methanol rinse, Br_2:CH_3OH polish, 30 minutes CH_3OH rinse, etch with HCl (10 Vol%):CH_3OH, methanol rinse 1 minute.

The surface states were studied up to some days after etch treatments by means of XPS measurements. Figure 11 shows the XPS binding energy spectra of P (2p) for case (a) and Figure 12 presents the change of binding energy of O (1 s) up to forming of the indium (III) orthophosphate. Figure 13 demonstrates the dependence of the relations of signal peak areas (their sums were normalized to be equal to one) c[PO_4]:c[P] on the time of atmospheric oxidation. Both functions (a) and (b) cut the lg t-axis at different time points. This time would be named "seed forming time" up to the begin of the indium (III)-orthophosphate phase formation. The changes of the binding energies of P (2p) and O (1s) levels in Figure 11 and Figure 12 permit to draw following conclusions of a model mechanism; respectively:

$$InP^{(m)} + H_2O \rightleftharpoons InP^{(m)} \times H_2O - \text{adsorpt}\,.$$

2. cracking of In–P bonds by catalytic H^+-ionic initiated hydrolysis

$$(InP)_x^{(S)}\ x\ H_2O \rightarrow (InP)_{x-1}^{(S)}\ (InOH)\ (PH)$$

3. Oxidation with O_2 from air and dehydratation of [InOH] accelerated by hygroscopic properties of developing hypophosphits, phosphits and hypophosphates. Spectra T and U sketched the reaction:

$$2[InOH] \rightarrow [In]\,[InO] + H_2O$$

correlated by the O (1 s) binding energy increasing at 530.0 eV.
4. Spectrum T in Figure 11 represents the mixture of different oxidation stages formed to $[In]\,[InO]\,[InOH]^{q+}\ [H_2PO_2]\,[HPO_3]\,[PO_4]^{q-}$ with the first oxide seeds of indium(III) orthophosphate phase.
5. Beside that reactions it may be happen another reactions step like this (spectra U—V, Figure 13)

$$InO \times HPO_3 \rightarrow InPO_4 + H^+$$

6. After forming of the indium(III)-orthophosphate phase it could not be observed further mixture of different oxides.

The stoichiometric relation of P:In approximates the value of 1.0 ± 0.02. At the end following structure is supposed to be present:

$$(In)_x\ (InO)_y\ (InOH)_z\ (PO_4)_{x+y/3+3z/2} \quad x \gg y + z$$

This investigations of atmospheric oxidation of (100) InP confirm the estimations of In–P–O phase diagrams made by SCHWARTZ et al. for room temperature too.

Conclusion

The following polishing treatment is useful to obtain optical smooth and oxide free (100)- and (111) P surfaces of InP:

- rinsing with trichlorethylen, aceton, methanol
- pre-etching with $(NH_4)_2S_2O_8:H_2SO_4:H_2O$ (15 g:75 g:15 g/RT/1 min.)
- rinsing with methanol
- Br_2–CH_3OH-polishing (1%/RT/1 min.)
- rinsing with methanol (90 min.)
- etching with HCl:CH_3OH (RT/10 s)
- rinsing with methanol.

The present investigations were done within the Arbeitsgemeinschaft $A^{III}B^{V}$-Halbleiter.

References

BERTRAND, P. A.: J. Vac. Sci. Technol. **18**, 28 (1981)
BINDUNGSENERGIE-KATALOG für das Gerät "Vacuum Generators ESCA III" (1972)
CHYE, P. E., SU, C. Y., LINDAU, I.: Surf. Sci. **88**, 439 (1979)
CLARK, D. T., FOX, T., ROBERTS, G. G., SYLKES, W. R.: Thin Solid Films **70**, 261 (1981)
FIEDLER, F., SCHLACHETZKY, A.: J. Cryst. Growth **53**, 623 (1981)
KAZMERSKI, L. L., IRELAND, P. J., SHELDON, P., CHU, T. L., CHU, S. S., LINH, C. L.: J. Vac. Sic. Technol. **17**, 1061 (1980)
KENZO, A., TOSHIRO, H., SATOSHI, S.: J. Cryst. Growth **46**, 73 (1979)
KÖNIG, U., KECK, W.: J. Less. Comm. Met. **90**, 299 (1983)
NAKAJIMA, K., YAMANAKI, S., AKITO, K.: J. Cryst. Growth **61**, 535 (1983)
NELSON, A., GEIB, K., WILMSEN, C. W.: J. Appl. Physics **54**, 4134 (1983)

Nelson, A. W., Westbrook, L. D., White, E. A. D.: J. Cryst. Growth **58**, 236 (1982)
Nishitani, Y., Kotani, T.: J. Electrochem. Soc. **126**, 2279 (1979)
Palme, H.: Z. Anorg. Chemie **112**, 97 (1920)
Phillips, J. C.: Rev. Modern Phys. **42**, 317 (1970)
Remy, H.: Lehrbuch der anorg. Chemie, Bd. 1, Leipzig 1970
Sing, S., Williams, R. S., van Uitert, L. G., Schlier, A., Camlibel, I., Bonner, W. A.: J. Electrochem. Soc. **129**, 447 (1982)
Schwartz, G. P., Sunder, W. A., Griffths, K. E.: Appl. Phys. Lett. **37**, 925 (1980)
Trapp, K. D. C., Eramanis, F. E.: J. Electrochem. Soc. **130**, 1381 (1983)
Wager, J. F., Ellsworth, D. L., Goodnick, S. M., Wilmsen, C. W.: J. Vac. Sci. Technol. **19**, 513 (1981)
Wager, J. F., Wilmsen, C. W.: J. Appl. Phys. **51**, 812 (1980)
Waldrop, J. R., Kowalczyk, S. P., Grant, R. W.: Appl. Phys. Lett. **42**, 454 (1983)
Wilmsen, C. W., Kee, R. W.: J. Vac. Sic. Technol. **15**, 1513 (1978)

(Received July 20, 1987)

Authors' address:

[illegible]. E. Kurth, [illegible]. A. Reif, Dr. V. Gottschalch, Dr. sc. J. Finster, Prof. Dr. sc. E. Butter
Sektion Chemie der Karl-Marx-Universität
Talstr. 35
Leipzig
DDR-7010

Cryst. Res. Technol.	23	1988	1	127–133

E. BUHRIG, K. HEIN, G. WALTER, B. GEIDEL

Bergakademie Freiberg

Calculation of One-Dimensional Temperatures Profiles in a Crystal Growth Furnace with Special Respect to Germanium

Dedicated to Professor Hermann NEELS on the occasion of his 75th birthday

In the paper a one-dimensional method of calculation for temperature profiles in crystal growth furnaces is described. It is characterized by high flexibility. The model applied to our own calculations is based on heat balances in segments of a multi-zone furnace and presupposes the availability of experimental data which allow the coefficients to be derived which are necessary to carry out the calculations. — The results of numerical calculations demonstrate the applicability of an existing furnace unit to directional crystallization of germanium.

Im Beitrag wird eine eindimensionale Berechnungsmethode für Temperaturprofile in Kristallzüchtungsöfen beschrieben. Sie ist durch hohe Flexibilität gekennzeichnet. — Die verwendete Methode basiert auf Wärmebilanzen in den Segmenten eines Vielzonenofens und benötigt experimentelle Daten zur Ermittlung der erforderlichen Koeffizienten. Die Resultate der numerischen Berechnungen demonstrieren die Anwendbarkeit eines Ofens zur gerichteten Kristallisation von Germanium.

1. Introduction

In the growth of single crystals the optimum technological parameters have to be adjusted with high accuracy. In addition to the chemical conditions and the geometry of apparatus, in particular, the thermal conditions must be considered. The horizontal Bridgman technique is characterized by that the crystallization from the melt is effected by a relative motion between sample and temperature profile, which can be realized by moving either the sample or the furnace or, when these are fixed, by a migrating temperature field. The temperature regime within the sample is considerably influenced by the local distribution of heating power in the furnace, the heat transfer parameters and the heat transport within the sample. When optimizing these parameters, emphasis is laid on the following requirements. An important requirement is to prevent an unwanted temperature change within the sample. To make this possible, an efficient temperature measuring and controlling technique must be available. The temperature constancy thus attainable in modern crystal-growing equipments is reported to be less than ± 0.1 K. In addition, it is to be avoided that, with a constant temperature on the furnace wall, deviations from the required value in the sample occur due to convection phenomena within the furnace chamber, changes in the emission and absorption coefficients etc. Additional requirements concern the local temperature distribution. In part, it is necessary that, in the furnace, zones with high temperature constancy (e.g., $dT/dx < 0.1$ K/cm) should be located immediately adjacent to zones with a steep temperature gradient (e.g., $dT/dx > 10$ K/cm). his is feasible only in a furnace of special design with a great number of separately controlled heating zones (BUHRIG et al.; PARSEY et al.).

The solution of these complex problems exclusively by experiments necessitates high expenditures. For this reason, the calculation of thermal conditions in developing crystal growth equipments and techniques may be an effective way to reduce the research and development expenditures. In the following, a simple method of mathematical modeling of one-dimensional temperature profiles is reported with special respect to growth of germanium crystals by the Bridgman technique.

2. Mathematical model

In the literature two basically different principles are used to calculate temperature profiles (CHIN et al.; NAUMANN, I and II; DEUS et al.). The analytical method applies closed functions that, as a rule, are relatively complicated, being able to be fitted to special and changing boundary conditions only in certain cases. On the other hand, the numerical methods of modeling the thermal conditions are essentially more flexible in considering various technological parameters. For these it is expedient to use computers.

The method applied to our own calculations is based on the heat balances in segments of a furnace model as shown in Figure 1, where the symbols have the following meaning:

n	consecutive number of segments $n = 0, N + 1$ denote the conditions, respectively, on the left-hand and right-hand side of the furnace; $n = 1 \dots N$ furnace segments
N	number of furnace segments
$T\,(1 \dots N)$	temperature to be calculated as a function of position in the furnace (°C)
$T\,(0, N + 1)$	temperature at the faces of the furnace (°C)
$P\,(0 \dots N + 1)$	heat transfer coefficients (W/K) between furnace segments and at the faces (in axial direction)
$PA\,(1 \dots N)$	heat transfer coefficient (W/K) outwards (in radial direction)
$TA\,(1 \dots N)$	external temperature (°C)
$\mathring{Q}\,(1 \dots N)$	heating power (W) supplied to the segments.

It is assumed that each segment, n, thermally interacts with segments $n - 1$ and $n + 1$ and with the environment ($TA(n)$) and that the heat $\mathring{Q}(n)$ is supplied or removed. From that the balance equation for segment n is obtained as follows:

$$\mathring{Q}(n) = P(n-1)\,\big(T(n) - T(n-1)\big) + P(n+1)\cdot\big(T(n) - T(n+1)\big) + \\ + PA(n)\,\big(T(n) - TA(n)\big) \quad (1)$$

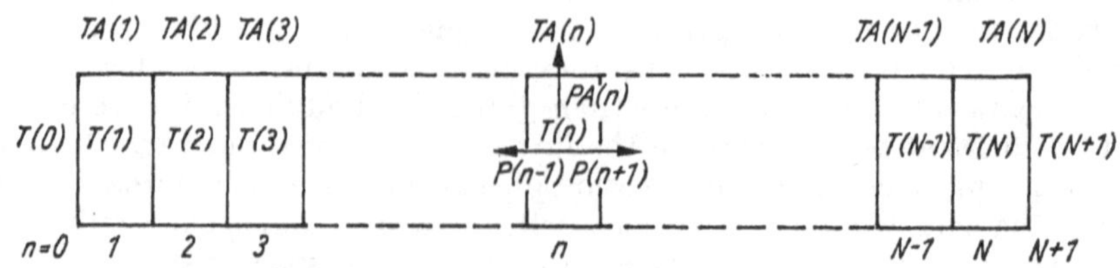

Fig. 1. Segmental model for establishing heat balances

If within the furnace the heat is transferred by radiation, an equation of the type

$$\mathring{Q}(n) = P'(n-1)\left(\frac{T(n)+T(n-1)}{2}\right)^3 (T(n) - T(n-1) +$$

$$+ P'(n+1)\left(\frac{T(n)+T(n+1)}{2}\right)^3 (T(n) - T(n+1) +$$

$$+ PA(n)\,(T(n) - TA(n)) \tag{2}$$

based on the Stefan-Boltzmann law, can be used as model base.

By formulating equations (1) and/or (2) for all segments, we obtain a system of equations the solution of which yields the unknown temperatures $T(n)$. For the solution a Gauss-Seidel iteration is very well suited due do its fast convergence. The iteration is broken off, when the differences between the i-th and $(i-1)$-th approximation are

$$[T_i(n) - T_{i-1}(n)]_{\max} < S\,. \tag{3}$$

If the break-off bound, S, is fixed low enough, the solution vector $T_i(n)$ describes the equilibrium state for the selected input quantities $T(0)$, $T(N+1)$, $TA(n)$, $P(0)$, $P(N+1)$, $P(n)$ or $P'(n)$ and $\mathring{Q}(n)$ with the values of $\mathring{Q}(n)$ characterizing a heating power constant in time.

Similarly to this approach it is also possible, by combination of equations (1) and (2), to take into consideration position- and temperature-dependent heat conduction and radiation amounts or to include more distant segments ($n \pm 2$, $n+3$ etc.) for the heat radiation.

In practice, controlled heaters are used in crystal rowth furnaces. This procedure also can qualitatively be simulated by the model descibed above. The controlled heater is characterized by that, at a defined point, n^*, the temperature is measured and compared with a required value $T^*(n^*)$. If $T(n^*)$ is different from $T^*(n^*)$, the heating power $\mathring{Q}$ must be changed. In the model this operation is performed by various modes based, for instance, on an on-off control or on a change in $\mathring{Q}$ which is proportional to the deviation from the required value. In the examples described below the adjustment of heating power is performed by correcting it after each iteration or after a selected number of approximations by the use of the following equation:

$$\mathring{Q}^* = Q(T^*(n^*)/T(n^*))^{\nu}\,. \tag{4}$$

Here also equation (3) can be used for the break-off criterion. The exponent ν must be chosen $\nu > 0$, causing a more or less quick approximation of $T(n^*)$ to $T^*(n^*)$. The minicomputer MC 80 and a personal computer 5120 were used for the numerical treatment of the model.

3. Model calculations on the directional crystallization of germanium

Aim of the calculations was to evaluate a horizontal tubular furnace with regard to its suitability for directional crystallization of germanium. The geometry of the model copy of the original furnace is shown in Figure 2. It exhibits 39 segments and 5 heating zones (2 unheated and 3 separately controllable zones). The thermostated average external temperature was fixed to be $TA = 910$ °C. By using appropriate input quantities a very good agreement between the measured profile (data points ● in Fig. 3) and the calculated profile (curve 1) could be attained, the residual deviations

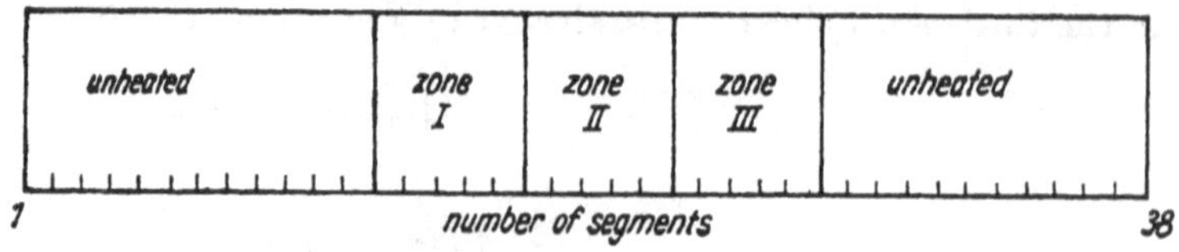

Fig. 2. Geometry of the furnace unit

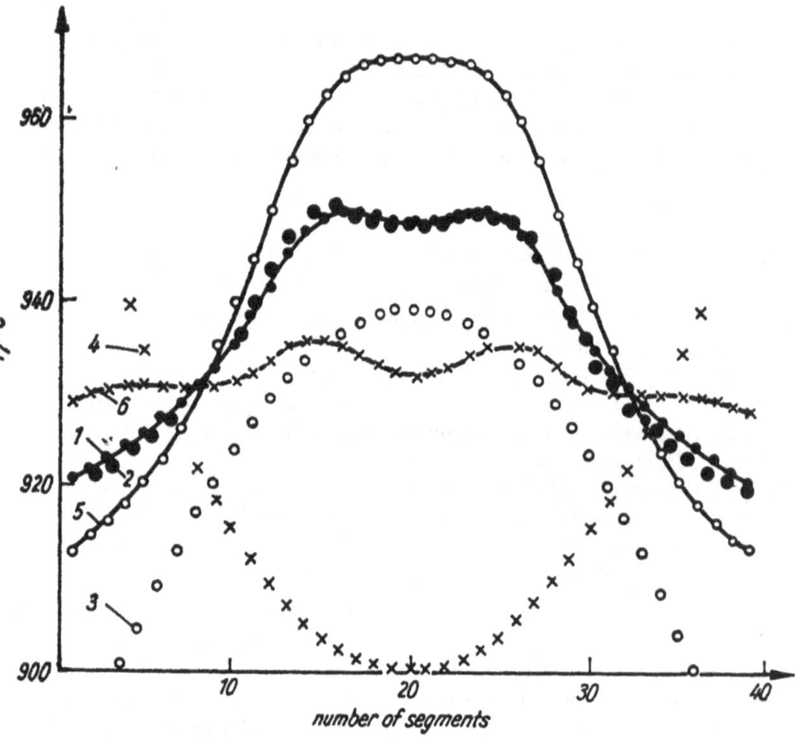

Fig. 3. Calculated and measured temperature profiles: 1 $T(n)$ (TA = 910 °C) (– . –), 2 T experimental (●), 3 $TA(n)$ (with maximum) (○), 4 $TA(n)$ (with minimum) (×), 5 $T(n)$ (TA with maximum) (– ○ –), 6 $T(n)$ (TA with minimum) (– × –)

being $\Delta T < 1$ K. It can be seen that under the given conditions germanium can be melted within 18 segments $(T(n) > 938\ °C)$.

The first problem to be solved arose from the question of how a position dependence of external temperature does affect the thermal profile (Fig. 3). For that, parabolic curves of the external temperature, TA, with maximum (curve 3) or with minimum (curve 4) were simulated. Curves 5 and 6 show the temperatures within the furnace. Whereas an external temperature profile with minimum is obviously unsuited to crystal growth, a profile of TA with maximum results in an extension of the range above $T_S = 938$ °C to 22 segments and in an increase of the gradient to ~ 5 K per segment, which must be considered a considerable benefit as compared with the profile at constant external temperature. However, also overheating is increased to $\Delta T_{Ü} \sim 20$ K. If we lower the temperature of the system down to $T_{max} \sim 950$ °C in order to avoid the higher overheating, the above-mentioned benefit disappears, the melting range being only 14 segments with a gradient of ~ 5 K per segment.

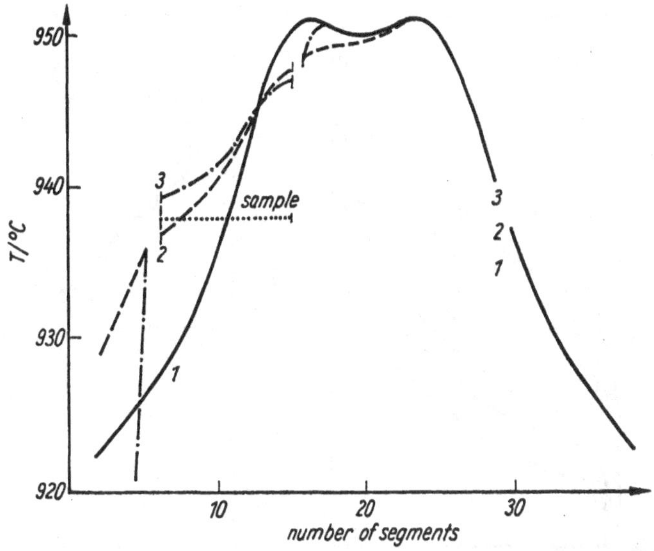

Fig. 4. Calculated temperature profiles $T(n)$: 1 without sample, 2 sample without heat insulation, 3 sample with heat insulation

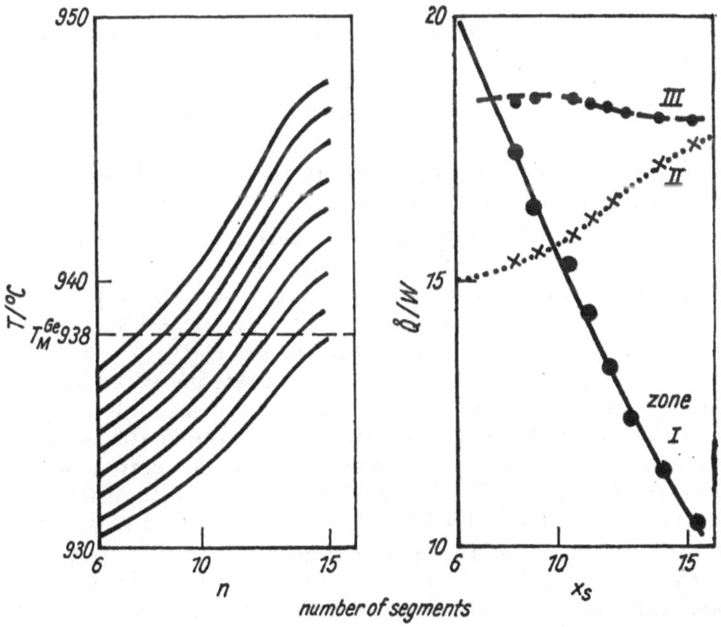

Fig. 5. Calculated temperature profiles and heating powers (stepwise lowering of $\dot{Q}$ in zone I, sample without heat insulation, x_s = position of interface)

The second problem was concerned with the thermal conditions within the furnace chamber in the presence of a sample. When the germanium sample sealed in a fused-silica ampoule is introduced into the furnace, the temperature profile effecting the crystallization is changed. Experiments have shown that, in particular, a flatter temperature gradient arises in the sample region. This phenomenon is shown in Fig-

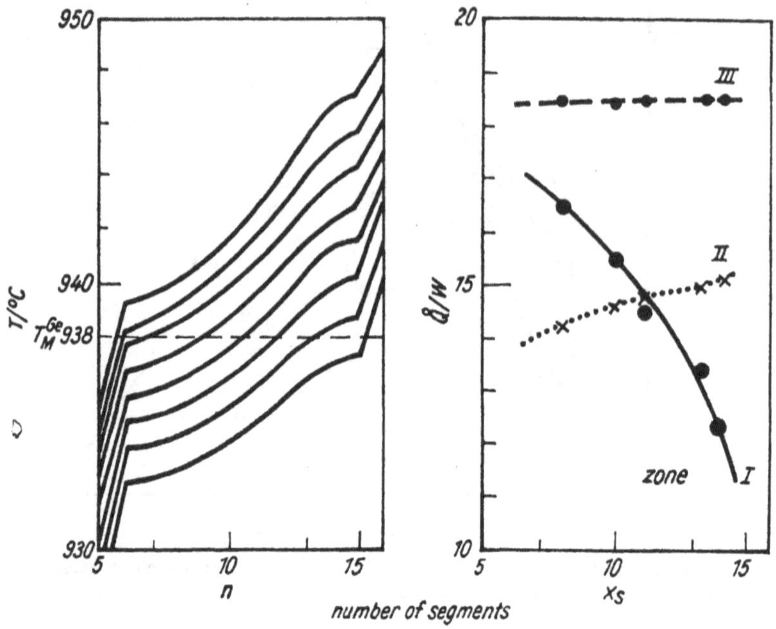

Fig. 6. Calculated temperature profiles and heating powers (stepwise lowering of $\mathring{Q}$ in zone I, sample with heat insulation, x_s = position of interface)

ure 4 by the calculated curves 2 and 3. Curve 3 represents the temperature change for simulated heat insulation at the ends of the sample.

Besides the flattening of the temperature gradient, a large local shift of the melting temperature can be found. Figures 5 and 6 show calculated temperature profiles of the germanium sample for the case of stepwise decreasing $\mathring{Q}$ in zone I and keeping the temperatures $T(n^*)$ in zones II and III constant by the use of equation (4). Migration of the melting temperature through the sample (gradient freezing) occurs irrespective of whether a heat insulation at the faces of the ampoule exists or not.

However, the temperature gradient within the system with heat insulation always exhibits the smaller value. This results in a growth rate increased by a factor of about 2. Attention must be given also to the differences in heating power programs, which result from the changed thermal properties of the sample. These calculated results can be transferred to a crystallization process only when a state near the equilibrium is always maintained by very low solidification rates. Under this condition a good qualitative agreement was obtained between calculated and measured temperatures in the model crucible and with the results of growth experiments.

A directional crystallization of the germanium sample could be realized under the measured and calculated conditions as represented in Figures 3–6. However, very high requirements as to the precision of the temperature measuring and controlling technique are derived from this temperature gradient of about 4 K/cm.

In order to position the interface between seed and melt at the beginning of the process with an accuracy of 1 mm, it is necessary to measure and adjust the absolute value of temperature within $\pm$ 0.2 K. If, in addition, we consider the fact that the average crystallization rate of $3 \cdot 10^{-4}$ cm/s should not vary by more than 10%, the temperature must be controlled in time with such an accuracy that it remains constant within $\sim 10^{-3}$ K/sec.

These data indicate the high expenditures required of the measuring and controlling technique, demonstrating that changes in technology and design of growing equipment can considerably influence the thermal conditions in the sample an, thus, the course of the process.

4. Summary

The calculation method described allows the thermal conditions in tubular furnaces used for crystal growth to be simulated by an one-dimensional approximation. It is characterized by low expenditures and high flexibility. The application of this method presupposes the availability of experimental data that allow the coefficients to be derived which are necessary to carry out the calculation. By the experimental fit of the parameters the agreement between measured and calculated data which is required to make statements as to the trend can be achieved despite the model being highly simplified.

The applicability of an existing furnace unit to directional crystallization of germanium was demonstrated.

References

Buhrig, E., Hein, K., Koi, H.: Bergakademie Freiberg, Wiss. Jahresbericht 1985, pp. 68–79

Chin, L.-Y., Carlson, F. M.: J. Cryst. Growth **62**, 561 (1983)

Deus, P., John, M., Krause, H., Sauermann, H.: Cryst. Res. Technol. (in preparation)

Naumann, R. J.: J. Cryst. Growth **58**, 554 (1982)

Naumann, R. J.: J. Cryst. Growth **59**, 569 (1982)

Parsey, J. M., Jr., Thiel, F. A.: J. Cryst. Growth **73**, 211 (1985)

(Received June 20, 1987)

Authors' address:

Dr. E. Buhrig, Prof. Dr. sc. K. Hein, [illegible], G. Walter, [illegible]. G. Geidel
Sektion Metallurgie und Werkstofftechnik der Bergakademie Freiberg
PSF 47
Freiberg
DDR-9200

Res. Cryst. Technol.	23	1988	1	135–140

H. Kresse, P. Rabenstein, H. Stettin, S. Diele, D. Demus, W. Weissflog

Martin-Luther-Universität Halle, Sektion Chemie, and
VEB Laborchemie Apolda, Betriebsteil Spezialchemie Leipzig

Liquid Crystalline Swallow-Tailed Compounds (II)

Dielectric Relaxation

Dedicated to Professor Hermann Neels on the occasion on his 75th birthday

Dielectric investigations in the kHz-range on swallow-tailed liquid crystals show a strong decrease of the dielectric constant as well as of the dielectric loss parallel to the nematic director with decreasing temperature. In the perpendicular direction no strong anomaly has been observed. By consideration of X-ray data of the S_A phase the dielectric results can be interpreted by the dipole-dipole interaction which is forced by the anisotropy of repulsive forces of the respective molecules.

Dielektrische Untersuchungen im kHz-Bereich an schwalbenschwanzartigen Flüssigkristallen zeigten sowohl einen starken Abfall der Dielektrizitätskonstante als auch des Verlustes parallel zum nematischer Direktor mit sinkender Temperatur. In senkrechter Richtung wurden keine stärkeren Anomalien beobachtet. Unter Berücksichtigung von Röntgenwerten in der S_A-Phase können die dielektrischen Ergebnisse durch die Dipol-Dipol-Wechselwirkung erklärt werden, die durch die Anisotropie der repulsiven Kräfte der entsprechenden Moleküle verstärkt wird.

1. Introduction

Due to the technical importance a lot of liquid crystalline compounds have been synthesized in the last years (Demus, Zaschke). There are also attempts to change the classical concept of the rodlike molecular shape. The synthesis of swallow-tailed molecules is one of these examples (Weissflog et al. 1984). Typical representatives are substituted benzylidene malonates:

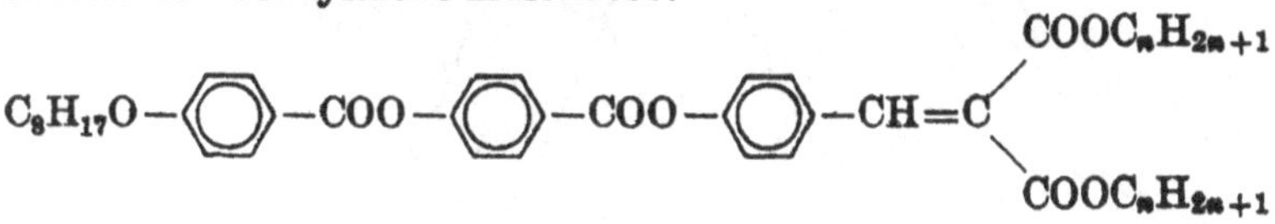

Subst.	n	Cr		S_A		N		is
1	5	·	327	·	360	·	383	·
2	10	·	336	·	357	·	364	·

X-ray studies performed at oriented samples of several members of the homologous series (Diele) prove a layer structure of the S_A type. The molecules are aligned parallel to the layer normal. The layer spacings d have been found to agree with the length L of the molecules in their fully extended form (Fig. 1). This agreement in connection

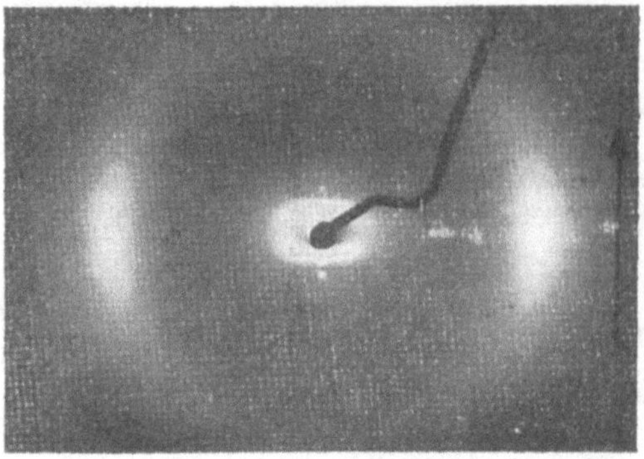

Fig. 1. Pattern of an oriented sample of 2 in the S_A phase ($T = 351$ K). The direction of the magnetic field is indicated by an arrow. $d = 4.55$ nm; $L = 4.65$ nm

with the strong anisotropy of the geometrical shape of the molecules led to a model with antiparallel packing of the molecules within the smectic layers (WEISSFLOG et al. 1984). The same results have been found by the aid of the static dielectric measurements (WEISSFLOG et al. 1985; KRESSE, RABENSTEIN). Dynamic dielectric investigations on the samples with different alkyl group length should give more information about the nature of this phenomenon.

2. Experiments

The dielectric investigations have been carried out in a microcapacitor (area 2 cm^2, distance 0.2 mm) by the aid of a modified Sawyer-Tower bridge. The substances have been oriented magnetically ($H \sim 0.8T$) parallel and perpendicular to the electrical measuring field. During the phase transition from the nematic (N) into the smectic A (S_A) phase

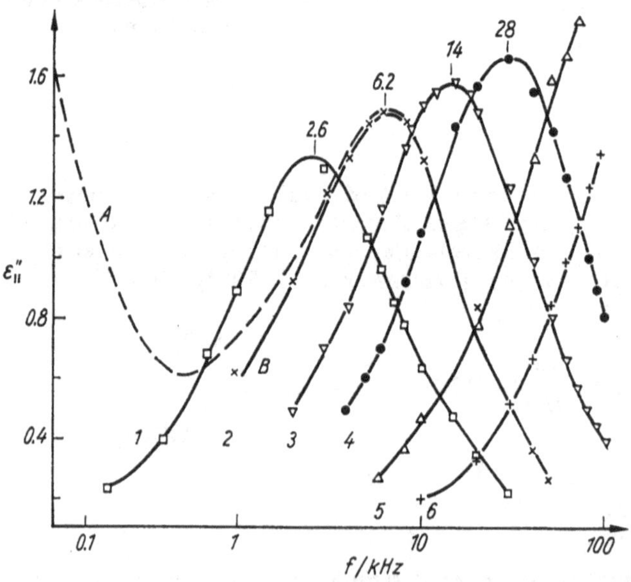

Fig. 2. Dielectric absorption curves of substance 1. Only for the second temperature the curves for the overall dielectric loss (A) and of the dielectric loss (B) are given. Numbers at the top: relaxation frequencies f_R/kHz. Temperatures: 1 322.4 K, 2 330.0 K, 3 338.2 K, 4 346.5 K (S_A phase), 5 360.2 K, and 6 367.5 K (N phase)

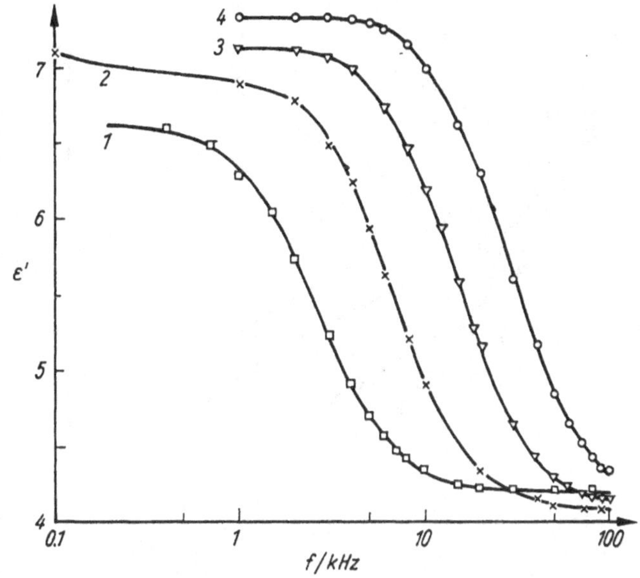

Fig. 3. Dispersion curves of substance 1. Temperatures see Figure 2

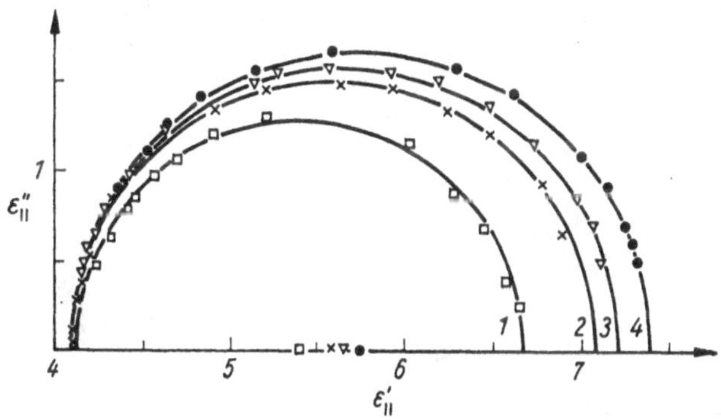

Fig. 4. Cole-Cole plot of the dielectric data of 1. Temperatures see Figure 2

the cooling rate was low (about 0.05 K · min^{-1}). Dielectric dispersion measurements have been done in the frequency range between 100 Hz and 100 kHz.

From these data the dielectric parameters $\varepsilon'_{\|}$, $\varepsilon''_{\|}$, and $\varepsilon'_{\perp}$ have been calculated at different frequencies. Experimental values of substance **1** are presented in Figures 2 and 3, resp. Dielectric absorption (Fig. 2) shows a strong increase at low frequencies due to the electrical conductivity (curve A) which can be easily substracted from the overall absorption in order to extract the dielectric loss (curve B). Electrical conductivity also is responsible for the formation of a double layer which gives an increase of the dielectric constant at low frequencies. Therefore, to avoid an error, the experimental dielectric constants at $f = 1$ kHz have been used for the calculation of static values, $\varepsilon_{\|01}$. The decrease of $\varepsilon'_{\|}$ due to the dielectric dispersion has been corrected using a Cole-Cole plot (Hill et al.) (Fig. 4). The high frequency limit in the parallel direction $\varepsilon_{\|02}$ has also been estimated and plotted in Figure 5. In the perpendicular direction no dielectric absorption in the investigated frequency range could be observed. In a similar way the dielectric data of substance **2** were calculated and plotted in Figures 5 and 6, resp.

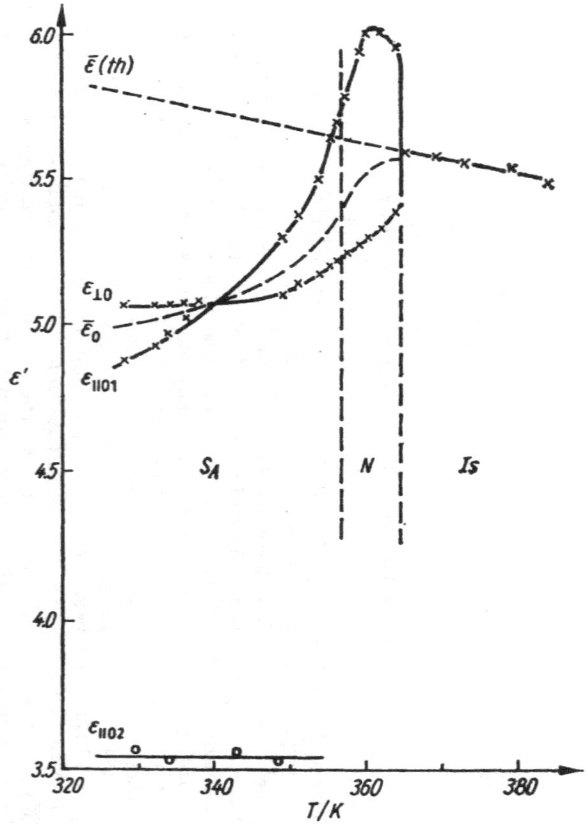

Fig. 5. Static dielectric constants of 1. $\varepsilon_{\parallel 01}$-static dielectric constant parallel to the director $\varepsilon_{\parallel 02}$-dielectric constant parallel to the director after finishing the kHz-absorption. $\bar{\varepsilon}_0$-mean value of the dielectric constant (calculated from the experimental data)

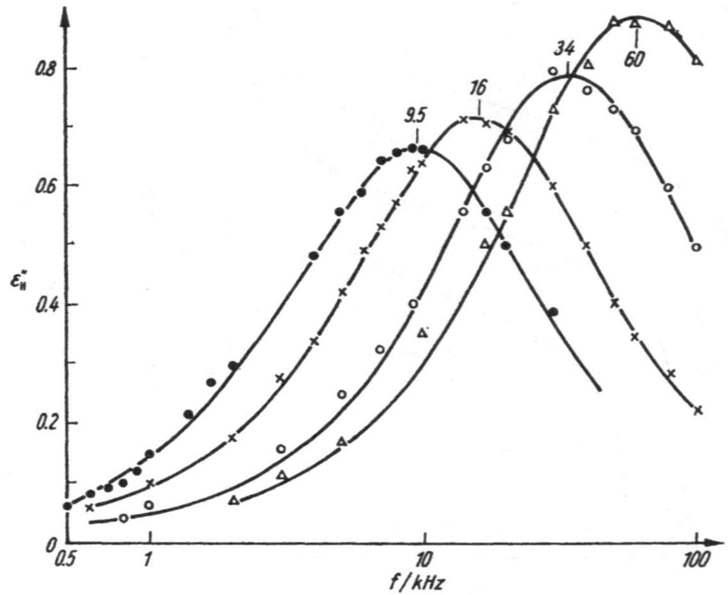

Fig. 6. Dielectric absorption curves of substance 2. Temperatures: 1 329.5 K, 2 334.0 K, 3 343.0 K, and 4 348.5 K

3. Discussion

The decrease of the dielectric constant $\varepsilon_{\|01}$ with decreasing temperature is a remarkable effect in this substance class. According to the MAIER-MEIER theory the dipolar part of the dielectric constant $\varepsilon_{\|01}$ is given by

$$\varepsilon_{\|01}(\mu) = \frac{NhF^2}{3\varepsilon KT}[\mu_1^2(1 + 2S) + \mu_2^2(1 - S)] + 1. \qquad (1)$$

If $\varepsilon_{\|01} - \varepsilon_{\perp 0} > 0$ then $\varepsilon_{\|01}$ should increase with decreasing temperature. The mean dielectric constant $\bar{\varepsilon}_0 = (\varepsilon_{\|01} + 2\varepsilon_{\perp 0})/3$ should also continuously increase in the same way. The comparison of model and experiment shows that only $\varepsilon_{\perp 0}$ exhibits the expected increase with increasing temperature. The introduction of a strong dipole-dipole correlation factor in equation (1) which is acting in the longitudinal direction (1) of the molecules

$$\mu_{1\text{eff}}^2 = g_1\mu_1^2 \qquad (2)$$

(DE JEU, BORDEWIJK) is a formal way for the description of this effect. Due to the hindrance in the reorientation as a result of an additional potential in the short range

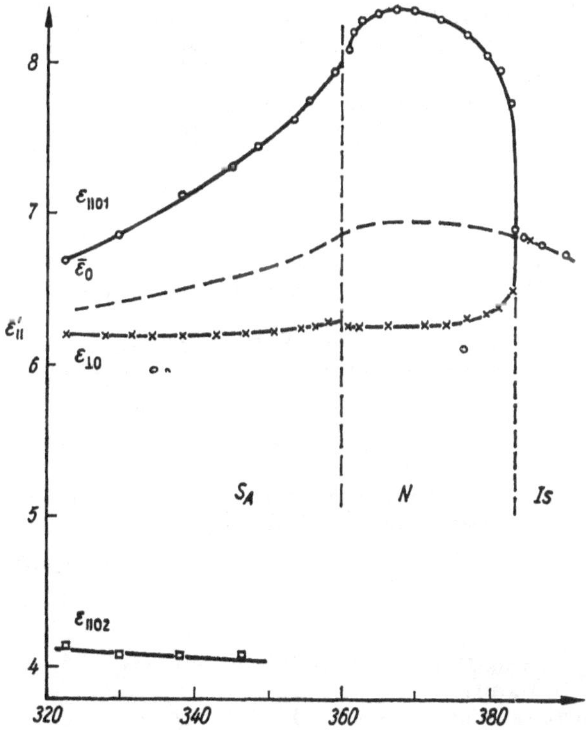

Fig. 7. Static dielectric constants of 2. $\bar{\varepsilon}$(th) expected mean value of the dielectric constant without taking into account the dipole-dipole correlation and the step of the density at the phase transition

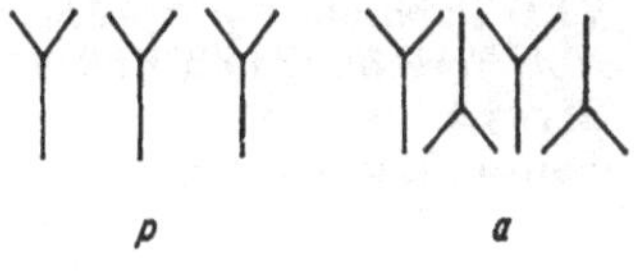

Fig. 8. Sketch of the parallel (p) and antiparallel (a) packing in a smectic A layer. Dee to a smaller free volume the antiparallel order of swallow-tailed molecules should be favoured

order, the effective dipole moment in the parallel direction is reduced. This effect can clearly be observed in the decrease of the absorption maxima of the low frequency absorption

$$\varepsilon''(\max) = \frac{NhF^2}{2\varepsilon} \frac{\mu_1^2 g_1}{3KT} (1 + 2S) \tag{3}$$

in Figures 2 and 6 resp. The constant values of $\varepsilon_{\parallel 02}$ indicate that the fast reorientation around the long molecular axis takes place and is not strongly influenced by the dipole-dipole interaction. For the same reason $\varepsilon_{\perp 0}$ shows normal behaviour according to the MAIER-MEIER model.

The unexpected high value of the dipole-dipole correlation should be connected with the geometrical shape of the molecules which is sketched in Figure 8. Only for the antiparallel orientation (a) we can expect a small free volume and therefore a stable phase. This effect can also be responsible for the preferred formation of smectic layers (WEISSFLOG et al. 1984). Therefore, we assume that the repulsive forces play an important role in the formation of the liquid crystalline state.

Geometrical reasons should also be responsible for the low relaxation frequencies of the substances for the reorientation around the short molecular axis. The activation energies according to the Arrhenius equation are 91.7 kJ mol^{-1} (1) and 90.3 kJ mol^{-2} (2), resp. In spite of the longer alkyl chain of 2 the relaxation frequency of this compound is about 1.4 times higher.

The presented results show the important role of geometrical factors in the formation of the nematic order.

References

DE JEU, W. H., BORDEWIJK, P.: J. chem. Phys. **68**, 109 (1978)

DEMUS, D., ZASCHKE, H.: Flüssige Kristalle in Tabellen II, Leipzig 1984

DIELE, S., MANKE, S.: in preparation

KRESSE, H., RABENSTEIN, P.: phys. stat. sol. (a) **100**, K83 (1987)

KRESSE, H., WEISSFLOG, W., NAGABHUSHAN, C., SHASHIDHAR, R.: phys. stat. sol. (a) **101**, K000 (1987)

MAIER, W., MEIER, G.: Z. Naturforsch. **16a**, 262 (1961)

WEISSFLOG, W., WIEGELEBEN, A., DIELE, S., DEMUS, D.: Cryst. Res. Technol. **19**, 583 (1984)

WEISSFLOG, W., DEMUS, D., KRESSE, H., STETTIN, H., RABENSTEIN, P.: Proc. 6th Liquid Crystal Conf. Socialist Countries, Halle 1985 (A 14)

(Received June 29, 1987)

Authors' addresses:

Dr. sc. H. KRESSE, Dipl. Chem. P. RABENSTEIN, Dr. H. STETTIN,
Dr. S. DIELE, Prof. Dr. D. DEMUS
Sektion Chemie der Martin-Luther-Universität Halle-Wittenberg
Mühlpforte 1
Halle/S
DDR-4020

Dr. W. WEISSFLOG
VEB Laborchemie Apolda, Betriebsteil Leipzig
Abt. Forschung und Entwicklung
Elsteraue 9
Leipzig-Lützschena
DDR-7143

Cryst. Res. Technol.	23	1988	1	K1–K5

Short Notes

K. FISCHER, E. SINN

Physikalisch-Technisches Institut der Akademie der Wissenschaften der DDR, Jena

The Correlation between the Crystallization Field of $Y_3Fe_5O_{12}$ and the Oxygen Ion Concentration of High-temperature Solvents

1. Introduction

For the preparation of yttrium iron garnet $Y_3Fe_5O_{12}$ (YIG) as bulk material or layer the composition of the high-temperature solvents (HTS) was varied in a wide range. Special compositions were selected with the aim to modify properties of the melts such as viscosity, density, surface tension, solubility, evaporation in thus manner that monocrystalline material is obtained with high yield and high quality. A series of investigations on phase relations in relevant systems was made to discover correlations between the crystallization fields of garnets on the one hand and the solvent compositions and the concentrations of the garnet forming oxides on the other hand (NIELSEN, DEARBORN; JONKER; ROBERTSON et al.; TIMOFEEVA, BYKOV).

In a previous paper (FISCHER, SINN) we reported the garnet formation as primary phase in different composed solvents and found correlations between the formed phases and the oxygen ion concentration (OIC) of these solvents. At these experiments the solute concentration, the ratio of the garnet-forming oxides, and the range of the crystallization temperature were standardized.

In practice of crystal growth and LPE experiments different solute concentrations and excesses of a garnet-forming component are used. We varied the concentrations of the soluted oxides in order to study the shift of the crystallization field and to check the correlation with the measured OIC of the mixture $PbO/PbF_2/B_2O_3/Fe_2O_3$. The experimental results are discussed in terms of ionic-particle models of the garnet formation.

2. Experimental and results

Experimental details were described in a previous paper (FISCHER, SINN). Now we studied the garnet formation at different solute concentrations included an excess of Fe_2O_3 in the solvent mixtures $(1 - x)$ PbO $\cdot$ $x(0.5\ B_2O_3 \cdot 0.5\ PbF_2)$ (a) and $(1 - x)$ PbO $\cdot$ x PbF_2 (b). These solvents represent parts of the solvent system $PbO/PbF_2/B_2O_3$.

In this paper the solute concentration C is defined as a product of molar fractions taking into consideration the cation ratio of the garnet phase and assuming the existence of particles with one trivalent cation (FISCHER)

$$C = (2[Y_2O_3]_{YIG})^{3/5} \cdot (2[Fe_2O_3]_{YIG} + 2[Fe_2O_3]_{ex})$$

where $[Y_2O_3]_{YIG}$: Y_2O_3 molar fraction for YIG formation,
$[Fe_2O_3]_{YIG}$: Fe_2O_3 molar fraction for YIG formation,
$[Fe_2O_3]_{ex}$: molar fraction of Fe_2O_3 excess.

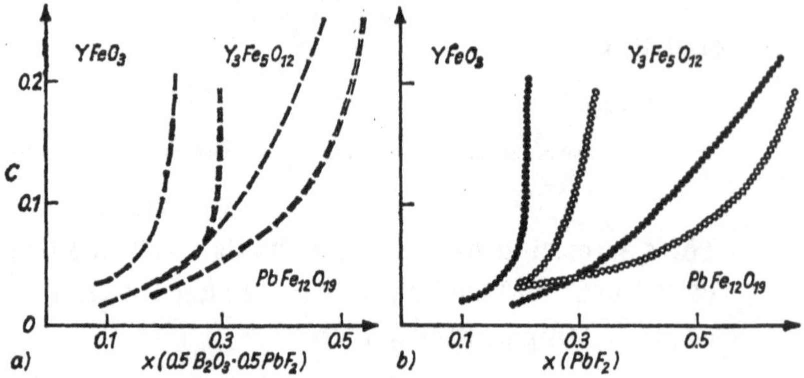

Fig. 1. Crystallization fields of YIG at different solute concentration C in the solvents $(1-x)$ $PbO \cdot x(0.5\ B_2O_3 \cdot 0.5\ PbF_2)$ (a) and $(1-x)\ PbO \cdot x\ PbF_2$ (b). = = =, ○○○ without Fe_2O_3 excess – – –, ●●● with Fe_2O_3 excess (0.9 solvent · 0.1 Fe_2O_3)

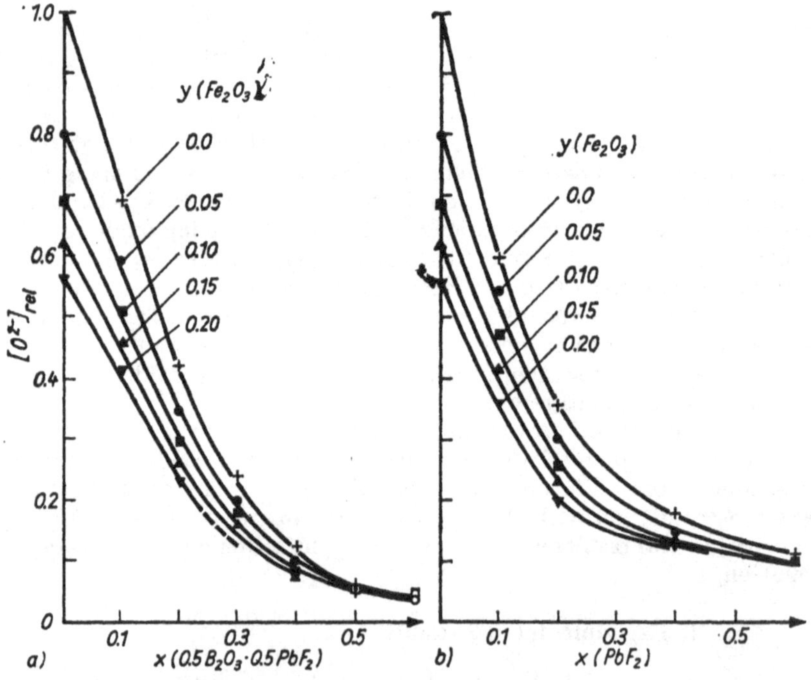

Fig. 2. Oxygen ion concentration of the solvents $(1-x)$ PbO · x $(0.5\ B_2O_3 \cdot 0.5\ PbF_2)$ (a) and $(1-x)$ PbO · x PbF_2 (b) with different contents of Fe_2O_3 ($y\ Fe_2O_3 \cdot (1-y)$ solvent)

In Figures 1a and 1b the concentration C is given as a function of the solvent composition. The lines demarcate the crystallization fields of the garnet phase from the adjacent phases $YFeO_3$ and $PbFe_{12}O_{19}$, respectively. As known the Fe_2O_3 excess effects the shifting of the garnet field; the garnet is formed in solvents with a higher PbO portion.

The OIC (LINZEN et al.) represents a characteristic property of solvents or solutions which contributes to the phase formation (FISCHER, SINN). Therefore we have measured the OIC of the used solvent mixtures after addition of Fe_2O_3. The results are shown in Figures 2a and 2b. It can be seen that the Fe_2O_3 added decreases the OIC of the pure solvent mixture. The complex ion FeO_4^{5-} (SUMITA et al.) is formed. The measure of decrease depends on the Fe_2O_3 concentration and the position of the chemical equilibrium $Fe^{3+} + 4\,O^{2-} \rightleftarrows FeO_4^{5-}$ (SINN et al.).

3. Discussion

We consider the formation of the garnet phase on the base of the reaction

$$3Y^{3+} + 2\,Fe^{3+} + 3\,FeO_4^{5-} \rightleftarrows (Y_3Fe_5O_{12})_{solid}$$

proposed by FISCHER and SINN. The solubility product for this reaction is given by the equation

$$L(T) = [Y^{3+}]^3\,[Fe^{3+}]^2\,[FeO_4^{5-}]^3\,.$$

The ratio $[Fe^{3+}]/[FeO_4^{5-}]$ is regulated by the OIC of the solution and the coefficient

$$K(T) = \frac{[FeO_4^{5-}]}{[Fe^{3+}]\,[O^{2-}]^4}$$

describes the position of the chemical equilibrium. With $[Y^{3+}] = 2[Y_2O_3]$ and $[Fe^{3+}] + [FeO_4^{5-}] = 2[Fe_2O_3]$ the general dependence of the solute concentration C on the solubility product $L(T)$, the equilibrium coefficient $K(T)$, and the OIC is obtained

$$C = (2[Y_2O_3])^{3/5} \cdot 2[Fe_2O_3]$$

$$= \sqrt[5]{L(T)\left(1 - \frac{[O^{2-}]^4}{K(T) + [O^{2-}]^4}\right)^{-2}\left(\frac{[O^{2-}]^4}{K(T) + [O^{2-}]^4}\right)^{-3}}$$

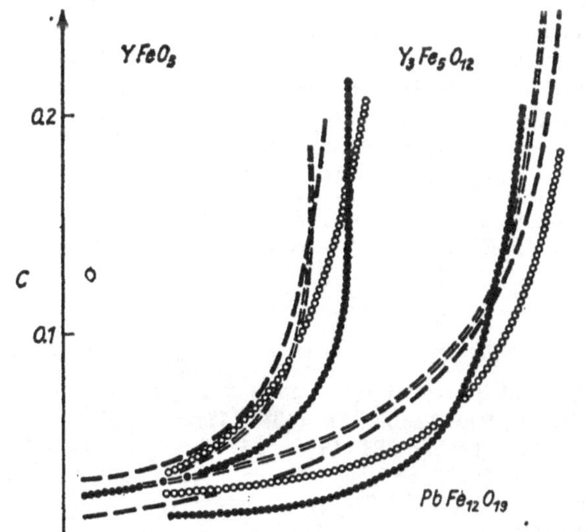

Fig. 3. Crystallization fields of YIG at different solute concentration C in dependence on the relative OIC of the used solvents including a Fe_2O_3 excess. (The symbols have the same meaning as in Figure 1).

For a given temperature the solute concentration C of the garnet-forming oxides is only dependent on the OIC of the high-temperature solution which is adjusted by the chemical substances and their concentrations.

Figure 3 shows the fields of the garnet crystallization in dependence on the concentration of the garnet-forming oxides C and the OIC of the different solvents including the Fe_2O_3 excess. A satisfactory agreement of the phase boundaries is observed. The influence of the soluted garnet on the OIC is neglected. The temperature is taken into consideration in the dependence of the solubility product $L(T)$ and of the equilibrium coefficient $K(T)$. The OIC is scarcely changed by the temperature in the interval for that matter (LINZEN).

The model used is able to give an interpretation of the observed experimental results. A dependence of solute concentration C on the OIC in the form

$$C = f(L(T), K(T)) \cdot g(\text{OIC})$$

can also be derived within other ionic models, e.g. publicated by TOLKSDORF, WELZ; VERTES; GÖRNERT; FISCHER, GÖRNERT; NITSCH et al.). The differences are found in the concrete dependence and shape of the terms $L(T)$, $K(T)$, and OIC.

The results are useful for a crystal grower. If the OIC of a solvent including an excess of a garnet-forming oxide is known a suitable composition of the high-temperature solution for the garnet formation can be chosen. Whithin the garnet crystallization field both the solute concentration with the crystallization temperature and physico-chemical properties of the solution can be varied.

4. Conclusions

The comparison of the crystallization fields of YIG in different composed HTS distinctly shows the correlation between the concentration of soluted oxides and the OIC. On the base of ionic formation mechanisms the dependence of the solute concentration on the OIC, on the solubility product, and on the equilibrium coefficients of co-operating reactions can be derived. Only the OIC is correlated with the composition whereas the solubility product and the equilibrium coefficients are independent of the composition of the high-temperature solution. At a given temperature the OIC essentially determines the kind of the crystallizing phases and its solubility.

The authors wish to thank Prof. Dr. F. Voigt for his stimulating discussions. Thanks are due to Mrs. R. Euler, A. Kupke, S. Manzel, and Mr. J. Voigt for technical assistance.

References

FISCHER, K.: Kristall und Technik **14** (1979) 835
FISCHER, K., GÖRNERT, P.: Cryst. Res. Technol. **17** (1982) 775
FISCHER, K., SINN, E.: Cryst. Res. Technol. **21** (1986) 711
GÖRNERT, P.: J. Cryst. Growth **52** (1981) 88
JONKER, H. D.: Cryst. Growth **27** (1975) 231
LINZEN, D.: private communication
LINZEN, D., SINN, E., FISCHER, K., KLUPSCH, T., BORNMANN, S.: Cryst. Res. Technol. **18** (1983) 165
NIELSEN, J. W., DEARBORN, E. F.: Phys. Chem. Solids **5** (1958) 202
NITSCH, K., MATYAS, M., BOK, J.: phys. stat. sol. (a) **98** (1986) 465
ROBERTSON, J. M., TOLKSDORF, W., JONKER, H. D.: J. Cryst. Growth **27** (1974) 241
SINN, E., LINZEN, D., FISCHER, K.: Cryst. Res. Technol. **20** (1985) 965

SUMITA, S., MORINAGA, K., YANAGASE, T.: Trans. Jap. Inst. Metals **24** (1983) 35
TIMOFEEVA, V. A., BYKOV, A. B.: Progr. Cryst. Growth Charact. **2** (1979) 377
TOLKSDORF, W., WELZ, F.: in: Crystal Growth, Properties and Application, ed. C. J. M. ROOIJMANS, Berlin-Heidelberg-New York 1978, p. 1
VERTES, A.: Acta Phys. Acad. Sci. Hung. **41** (1979) 209

(Received July 13, 1987)

Author's address:

Dr. rer. nat. Klaus FISCHER, Dr.-Ing. Ekkehard SINN
Akademie der Wissenschaften der DDR, Physikalisch-Technisches Institut
Helmholtzweg 4
Jena
6900

Cryst. Res. Technol.	23	1988	1	K6–K10

E. POLLERT, J. HEJTMÁNEK, K. JUREK, J. KAMARÁD, A. TŘÍSKA, D. ZEMANOVÁ

Institute of Physics, Czechoslovak Academy of Sciences, Prague

Effect of the Pressing on the Properties of the Superconducting $YBa_2Cu_3O_{7-\gamma}$ Phase

Introduction

Since the discovery of the superconducting behaviour in the Ba–Y–Cu–O system (CHU et al.; HOR et al. 1987a) enormous number of works devoted above all to the compound $YBa_2Cu_3O_{7-\gamma}$ appeared. Thus its structure (CAVA et al.; CAPPONI et al.; JIRÁK et al.) and effect of various substitution were studied (HOR et al. 1987b); WILLIS et al.).

Recently we published a short comment to the relation between the condition of preparation and the resulting electrical properties of the compound (POLLERT et al.). Now we report an extended study of the problem concerning the influence of the compacting pressure.

Preparation of samples

The starting materials, Y_2O_3, $BaCO_3$, and CuO, tested by chemical analysis were weighted and homogenized in the required ratio 1:2:2. preheated 12 hours at 1170 K under oxygen and once again homogenized. Then the series of pellets prepared by the compacting pressure ranging from 0.34 GPa to 4 GPa were heated at 1223 K under oxygen atmosphere for 70 hours and cooled down to room temperature by the rate of 3 K/min.

Methods of characterization

Phase composition and the lattice parameters of the $YBa_2Cu_3O_{7-\gamma}$ compound were determined by the X-ray diffractometer DRON-1, the CuKα radiation filtered by nickel filter was employed.

Classical GORTER's method was modified to determine the oxygen stoichiometry, i.e. concentration of Cu^{3+} ions (ZEMANOVÁ).

Investigation of the microstructure was done by means of scanning electron microscopy and the standard four-probe method was used for the temperature-resisting measurements.

Results

All prepared samples were found to be single phase with the lattice parameters $a = 3.825$, $b = 3.882$, $c = 11.659$ (space group Pmmm). They exhibit a metallic behaviour in the temperature range of 300 K $\div$ T^c_{onset} and subsequent sharp transi-

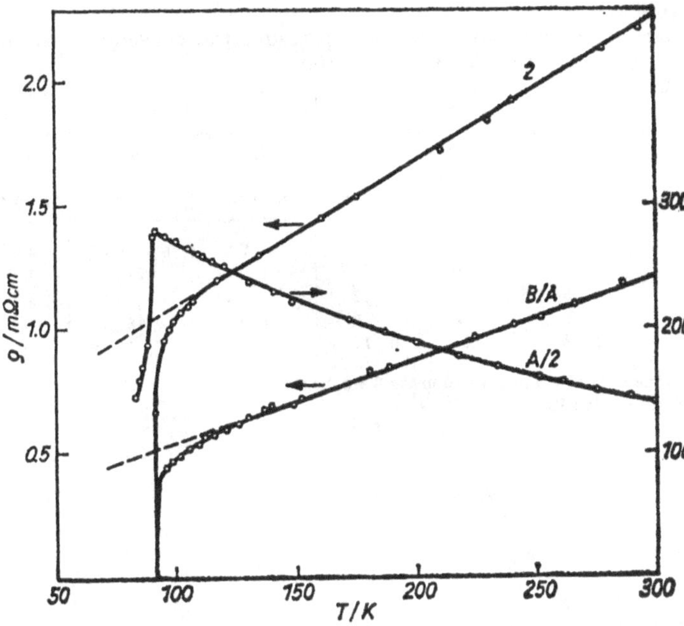

Fig. 1. Examples of the temperature dependence of the electrical resistance, notation see Tables 1 and 2

Table 1
Characteristic data of the prepared samples

No.	compacting pressure GPa	ϱ_{300} m Ω cm	ϱ_{95} m Ω cm	$Cu^{3+}/(Cu^{2+} + Cu^{3+})$ %
1.	0.34	2.51	1.129	6.1
2.	2.42	2.20	0.95	6.6
3.	0.9	1.6	0.681	6.2
4.	2	1.57	0.651	4.9
5.	4	1.76	0.717	7.4

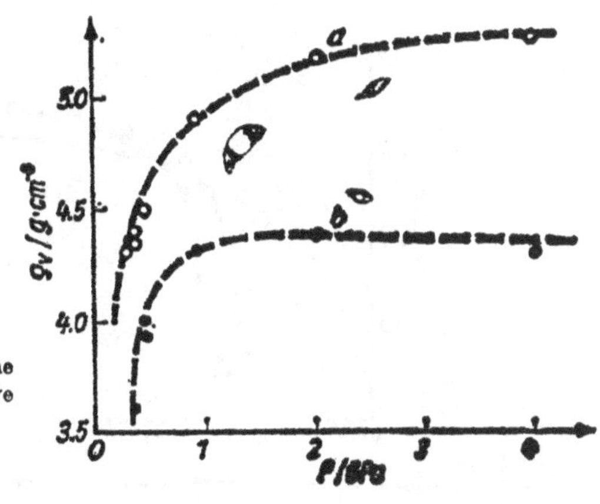

Fig. 2. Influence of the compacting pressure on the density of samples, curve a – before heating, curve b – after heating

Table 2
Effect of the additional isostatic pressing and subsequent thermal treatment on the density and electric properties

Sample	treatment				resistivity		density	
	pressure GPa	temperature K	atmosphere	time hours	ϱ_{300} m Ω cm	ϱ_{0s} m Ω cm	ϱ_v g/cm³	$\varrho_v/(\varrho_v)_t$ *) %
2	0	1223	O_2	72	2.2	0.95	3.9	62
A/2	1	300	air		140	275	4.7	73
B/A	0	1223 **)	O_2	72	1.21	0.42	5.6	88
C/2	3	823	air	0.5	110	410	6.2	98
D/C	0	1223	O_2	72	2.43	1.14		

*) Related to the theoretical value of the density 6.3 g/cm³
**) For about 1 hour at 1263 K

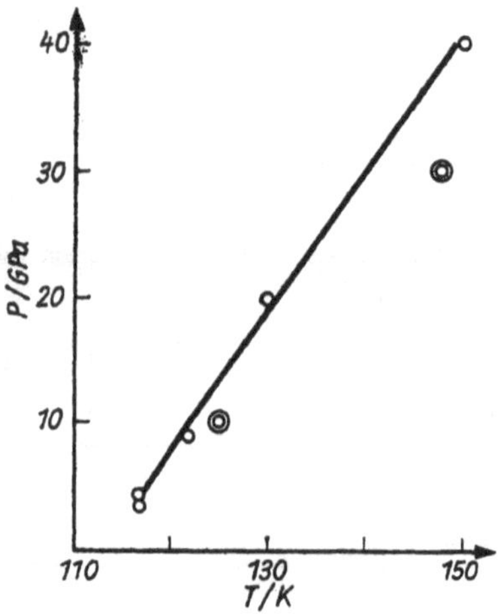

Fig. 3. Influence of the compacting pressure on the value of T^c_{onset}, ⊙ – indicate T^c_{onset} of the samples after the additional treatment

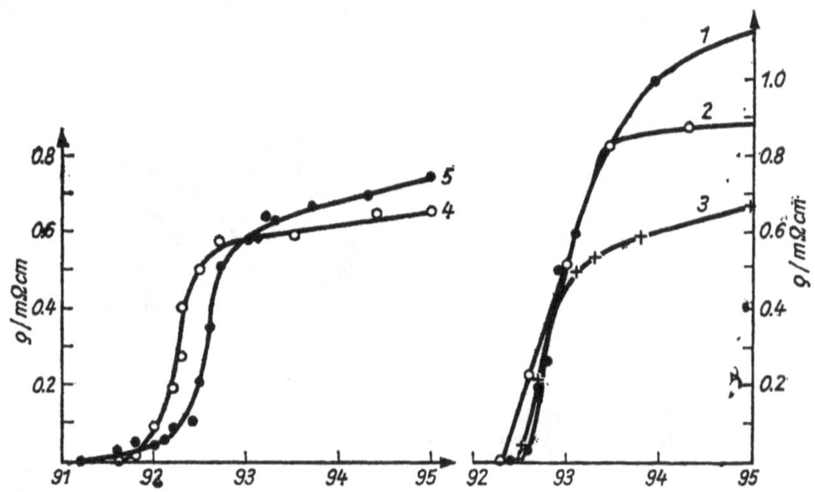

Fig. 4. Metallic – superconducting transition for the samples 1, 2, 3, 4, 5 (see Tab. 1).

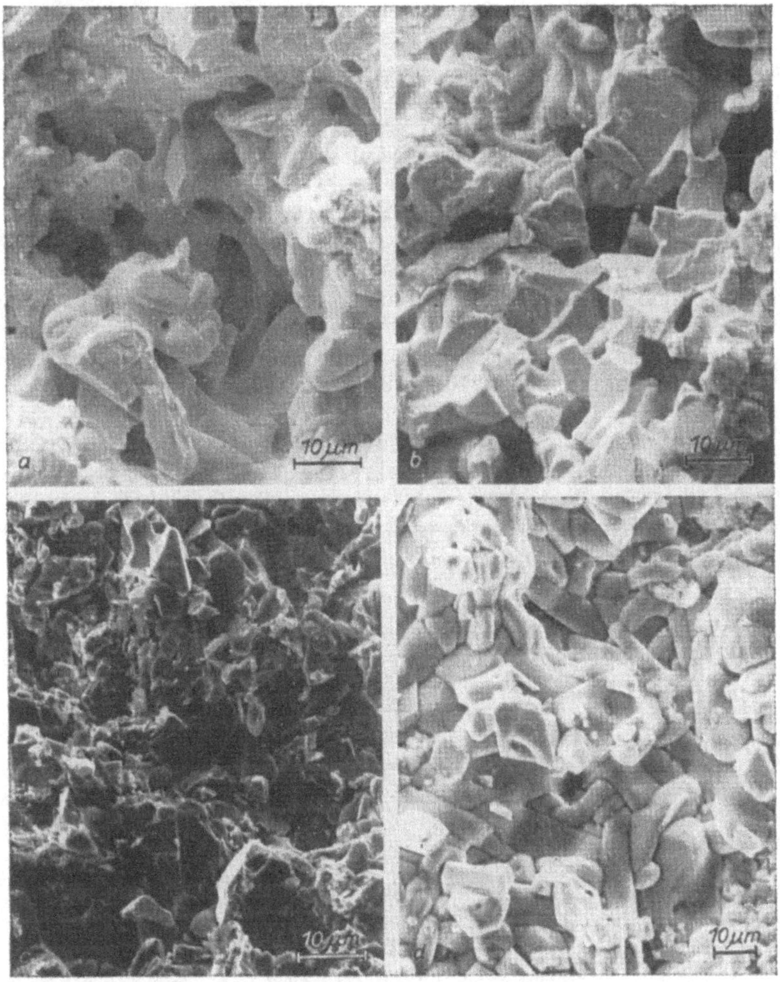

Fig. 5. Microstructure of the sample (2), a – surface of the cutted sample (2), b – broken sample (2), c – isostatically pressed sample (A/2), bridges between grains are broken, d – surface of the annealed sample (B/A); small particles on the surface are a phase separated from the matrix of the $YBa_2Cu_3O_{7-y}$-phase, probably Y_2BaCuO_5, not being observed in the interior of the sample. – indicates 10.00 µm

tion to superconducting state $\Delta T_c \sim 1$ K, examples are given in the Figure 1 and some characteristic data are summarized in Table 1. One can observe following effects of the compacting pressure:

– Its increase above 1 GPa had only a slight influence on the density of the samples and on the value of ϱ_{300} respectively. Significant lowering of the density after the second heating run, see Figure 2, is a consequence of the incomplete reaction during the calcination period, leading to the simultaneous formation of the phase $YBa_2 \cdot Cu_3O_{7-y}$ and unidentified intermediate products. Let us note that no nondecomposed carbonates were indicated by X-ray analysis as well as by the chemical test.

– Increase of the compacting pressure leads to an increase of the value T^c_{onset}, but simultaneously T^c_0 is slightly, about 1 K, decreased (see Figs 3, 4).

Existence of partially melted grains, joined by bridges and large cavities are typical for the microstructure of the prepared samples (see Figs 5a, b). Its formation was conditioned by a choice of the sintering temperature close to the peritectic one (ZEMANOVÁ).

Due to the equal heating conditions nearly no variation of Cu^{3+}/Cu^{2+} was found in the studied samples. Thus the influence of the oxygen stoichiometry on the resulting behaviour could not be observed.

The experiments listed in the Table 2 illustrate an effect of the additional isostatic pressing and thermal treatment on the density and electrical properties of the sintered samples. They both cause and increase of the density, but the pressing itself lead to a semiconducting behaviour of the prepared samples. Reason is in the character of the microstructure, existence of already described bridges between the grains. If a pressure is applied they are broken and can be restored by a thermal treatment at a suitable temperature about 1220 K, close to the mentioned peritectic point (see Figs 5, c, d).

Additional pressing influences the value of T^c_{onset} by a similar way as it was found for compacting pressure before the sinterign. Thus the original value of $T^c_{onset} = 117$ K ($P = 0.42$ GPa) increases to the value of 125 K (1 GPa) and 147 K (3 GPa) respectively (see Fig. 3). We cannot give any explanation of the surprising behaviour which will be studied in more detail.

References

CAPPONI, J. J., CHAILLOUT, C., HEWAT, A. W., LEJAY, P., MAREZIO, M., NGUYEN, N., RAVEAU, B., SOUBEYROUX, J. L., THOLENCE, J. L., TOURNIER, R.: submitted to Europhysics Letters

CAVA, R. J., BATTLEGG, B., VAN DOZER, R. B., MURPHY, D. W., SUNSHINE, S., SIEGRIST, T., REMEIKA, J. P., RIETMAN, E. A., ZAHURAK, S., ESPINOSA, G. P.: Phys. Rev. Lett., to be published

CHU, C. W., HOR, P. H., MENG, R. L., GAO, L., HUANG, Z. J., WANG, Y. Q., HUANG, C. Y.: Phys. Rev. Lett. 58 (1987) 908

GORTER, E. W.: Philips Res. Rept. 9 (3954) 1954

HOR, P. H., GAO, L., MENG, R. L., HUANG, Z. J., WANG, Y. Q., FOSTER, K.; VASSILIOUS, J., CHU, C. W., WU, M. K., ASHBURN, J. R., TORNG, C. J.: Phys. Rev. Lett. 58 (1987a) 911

HOR, P. H., MENG, R. L., WANG, Y. Q., GAO, L., HUANG, Z. J., BECHTOLD, J., FOSTER, K., CHU, C. W.: Phys. Rev. Lett. 58 (1987b) 1891

JIRÁK, Z., POLLERT, E., TŘÍSKA, A., VRATISLAV, Z.: submitted to phys. stat. sol.

NEVŘIVA, M., POLLERT, E., ŠESTÁK, J., TŘÍSTA, A.: submitted to Thermochimica Acta

POLLERT, E., HEJTMÁNEK, J., ZEMANOVÁ, D.: Czech. J. Phys. B37 (1987) 657

WILLIS, J. O., FISH, Z., THOMPSON, J. D., CHEONG, S. W., AIKIN, R. M., SMITH, J. L., ZIRNGIEBL, E.: submitted to J. Mag. Mag. Mat.

ZEMANOVÁ, D.: will be published

(Received July 15, 1987)

Author's address:

c/o E. POLLERT
Institute of Physics, Czechoslovak Academy of Sciences
Na Slovance 2
1840 40 Prague 8 — Libeň
Czechoslovakia

Cryst. Res. Technol.	23	1988	1	K11–K15

L. D. PRAMATAROVA, E. B. SAVOVA, G. M. MINCHEV, M. G. MIHAILOV

Institute of Solid State Physics, Bulgarian Academy of Sciences, Sofia, Bulgaria

Preparation of GaAs Substrates for MBE

1. Introduction

The process of cleaning and chemical treatment of the substrates prior to the epitaxial growth is important for the preparation of high-quality epitaxial layers by MBE. GaAs is successfully used as a substrate material for AlAs–GaAs superlattices as well as for GaAs–AlGaAs double heterostructure lasers (CASEY, PANISH). Besides this, it is suitable also as a substrate for the preparation of some quaternary compounds used in the production of optoelectronic devices.

Only a very small quantity of material is removed from the surface of the substrate in the vacuum chamber prior to the epitaxial layer growth by the method of MBE. Therefore, an inappropriate preliminary preparation of the substrate can lead to a poor quality of the grown GaAs epitaxial layers.

The aim of the present work is to study the influence of some systems of etchants and of the etching conditions on the surface morphology and the composition of the doped Sn–GaAs (100) substrates. The GaAs substrate surface was monitored with the help of "C–Pt" replicas, "Talystape", and XPS.

2. Experimental

Sn-doped (100) GaAs substrates were used with a free carrier concentrations of $(1.1-2.4) \cdot 10^{18}\,cm^{-3}$.

The surface of the substrates was treated in the following way:

a) The preliminary preparation of the substrates included in some experiments chemical-mechanical treatment and in others – mechanical polishing of the substrates originally with 28.0/μm (SiC) and then successively with 5.0/μm, 0.3/μm and 0.05/μm (Al_2O_3).

b) Cleahing of the substrates in boiling trichlorethylene (CCl_4) and acetone with following rinsing in bidistilled water.

c) Etching in concentrated HCl for 5 minutes with following rinsing in bidistilled water.

d) Chemical etching in:

I – $H_2SO_4/H_2O_2/H_2O = 3:1:1$ for 1 minute or for 5 minutes
II – only in concentrated H_2SO_4 for 1 minute
III – 3% Br_2/CH_3OH for 1 minute or for 5 minutes
IV – NH_4OH (25% solution)/H_2O_2 (30% solution)/H_2O_2 = 2:5:3 for 5, 10 or 14 minutes.

Immediately after etching, the substrates were rinsed in bidistilled water and then dried in air.

The etchants were always fresh and mixed in a magnetic agitator. The solutions were mixed and left to mature (from 30 minutes to 5 hours) to ensure a more reproducible oxidation activity. The etching was always carried out at room temperature.

3. Results and discussions

The system $H_2SO_4/H_2O_2/H_2O$ (I) is one of the most commonly used etchants for GaAs polishing. The surface of a chemical-mechanically (CM) treated (100) GaAs substrate is shown in Figure 1 and the surface of a (100) GaAs substrate (CM) treated and etched for 1 minute in etchant I, is shown in Figure 2. In both cases before etching the substrates, they were treated in concentrated HCl for 5 minutes. The micromorphology of the surface of the chemical-mechanically treated and etched in HCl substrates

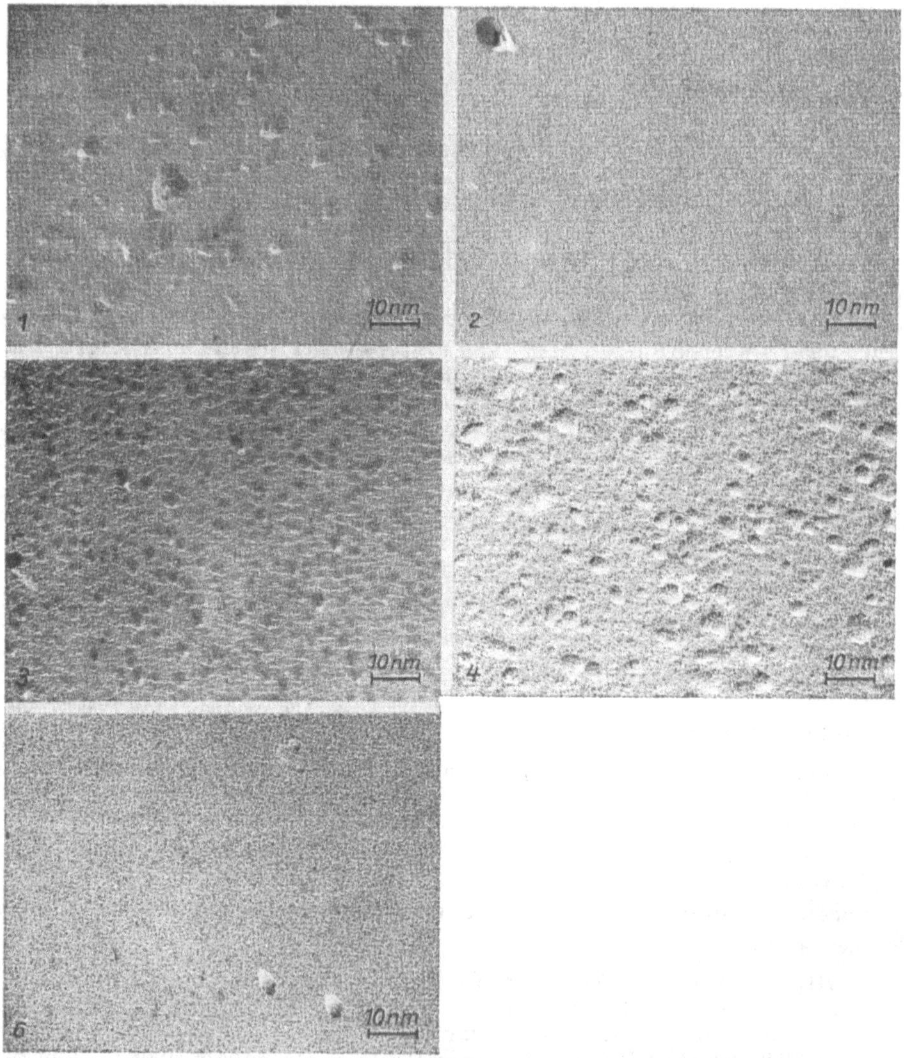

Fig. 1. Morphology of a chemical-mechanically treated (100) GaAs sample

Fig. 2. Morphology of (100) GaAs surface etched in $H_2SO_4/H_2O_2/H_2O$ = 3:1:1 for 1 minute

Fig. 3. Morphology of etched (100) GaAs surface, when the water content of etchant I is increased

Fig. 4. Morphology of etched (100) GaAs surface, treated only with concentrated H_2SO_4

Fig. 6. Morphology of etched (100) GaAs surface, treated in 3 % Br_2/CH_3OH for 1 minute

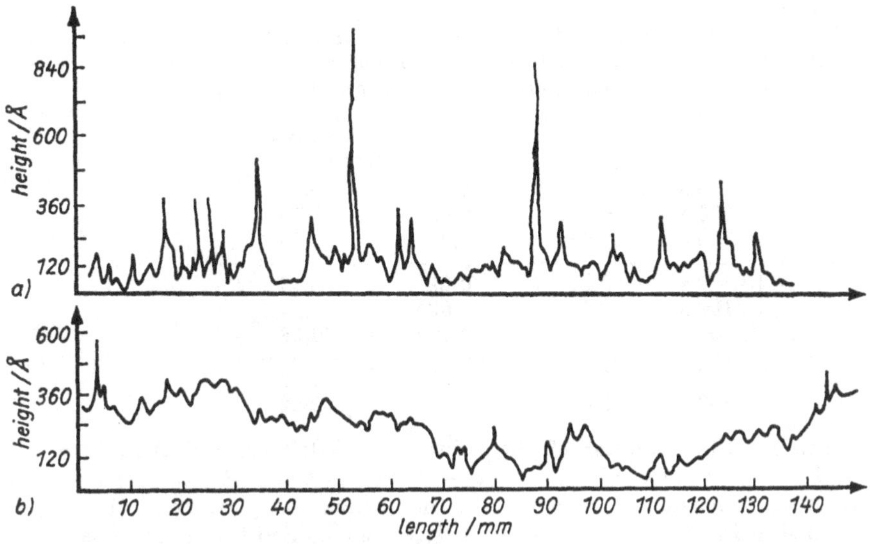

Fig. 5. Surface profiles of a (100) GaAs substrate measured with a Talystep a) before b) after treatment in $H_2SO_4/H_2O_2/H_2O$ = 3·1:1 for 1 min

nucludes a sufficiently homogeneous distribution of columns of identical dimensions: their diameter is about 350 Å, and their average surface density is $\sim 1.5\cdot10^{10}$ cm^{-2}. After etching with etchant I (Fig. 2), the surface appears smooth and the distribution of columns is more or less homogeneous.

When the water content of etchant I is increased, the morphology of the etched surface becomes more complex (Fig. 3).

And after treatment only with concentrated H_2SO_4 the whole surface is covered with serrations and the set of small and large columns becomes more complex (Fig. 4).

The surface of the samples treated in $H_2SO_4/H_2O_2/H_2O$ = 3:1:1 (etchant I) for one minute was studied with the help of "Talystape" (Fig. 5). Before chemical treatment the height of the serration was $\sim 0.064/\mu m$ and after chemical treatment for 1 minute $\sim 0.048/\mu m$.

The mean distance between columns, the columns increases 1.5 times and for longer times of etching, the height of the peaks becomes $\sim 0.023/\mu m$, i.e. the relief is smoothened.

In spite of the fact that $H_2SO_4/H_2O_2/H_2O$ etchants yield good surfaces, their etching rates for GaAs is not uniform over the substrate area: the so called "slopes" are formed at the edges.

That is why in this case it is better to use, for example, Br_2/CH_3OH etchant which etches GaAs quickly and is suitable for polishing and for achieving a shining surface. 1%–3% Br_2/CH_3OH is commonly used. This system, however, has a serious demerit: the great volatility of Br_2 leads to highly irreproducible etching rates.

The results of the electron microscope studies using the method of "C–Pt" replicas showed that the surface of the samples treated in 3% Br_2/CH_3OH for 1 minute was better then they were kept in the etchant for 30 minutes (Fig. 6) than when they were kept in it for 5 hours. The results obtained with the help of the "Talystape" showed that the surfaces of the chemical-mechanically treated substrates had micro-roughnesses with height $\sim 0.061/\mu m$, while after chemical treatment in 3% Br_2/CH_3OH for 1 minute, this value $\sim 0.053/\mu m$. In this case, it is characteristic that after

Table 1
XPS studies of (100) GaAs samples treated in $NH_4OH/H_2O_2/H_2O$ for 15 min with different preliminary treatments: a) chemical-mechanical + degreasing; b) polishing + degreasing; c) polishing without degreasing

	time	C/Ga	O/Ga	As/Ga
a	0 s	0.54	0.76	0.93
	200 s	0.046	0.033	0.78
	300 s	0.042	0.023	0.75
b	0 s	0.69	0.69	0.97
	100 s	0.044	0.035	0.67
	200 s	0.044	0.028	0.58
c	0 s	1.09	0.87	0.98
	200 s	0.015	0.025	0.62

1 minute treatment the diameter of the columns is 0.062/μm and their height is 0.049/μm, and after 5 minutes of treatment — 0.084/μm and 0.056/μm, respectively.

Using the method of the "C–Pt replicas", the surface morphology of (100) GaAs substrates etched in IV etchant $NH_4OH/H_2O_2/H_2O$ was studied too. This etchant possesses the useful ability to degrease the surface of the substrates as well as to promote the formation of complexes of ionized impurities Cu^2, Cr^2, Al^2 etc. which are adsorbed on the substrates' surface.

A study was carried out of substrates which were chemically-mechanically treated or polished and then etched in etchant IV for 5 minutes. Comparing the surfaces of the two substrates one can conclude that the surface of the chemically-mechanically treated samples is better with dimensions of the columns after etching: diameter 0.032/μm and height ~ 0.014/μm.

The surface relief after treatment in dilute solution of ammonia appears the same as after treatment in dilute sulphuric acid etchant.

After chemical treatment the surface of all semiconducting compounds is chemically active. They readily absorb carbon and other chemical elements from the atmosphere. Because of that, we carried out method of XPS studies of some samples treated in IV etchant but with different preliminary treatment:
a) CM + degreasing; b) polishing + degreasing; c) polishing without degreasing. The results of these studies are shown in Table 1. In all three cases of different treatment the ratio O/Ga on the surface remains approximatively one and the same (0.69—0.87) while deep in the volume after 200 seconds of Ar^2 ion bombardment this ratio is reduced to (0.038 — 0.025).

The lowest values of the ratio C/Ga were obtained for samples which were initially polished, and then degreased and etched for 15 minutes in IV etchant.

4. Conclusions

High quality surface samples were obtained by treatment in ammonia etchant — the roughness of the surface assessed with the help of the method of the "C–Pt replicas" and "Talystape" is about 0.014 μm while after treatment in 3% Br_2/CH_3OH it is about 0.053/μm and in $H_2SO_4/H_2O_2/H_2O = 3:1:1$ it is about 0.23/μm.

As with MOLODJAKOVA and ANDREEV the different surface morphology can apparently be connected with variations in the etching process sensitivity to defects in substrates, activated to different degrees, depending on the etchant composition and dilution.

Treatment of the samples only in concentrated H_2SO_4 which is suggested by some authors spoils the surface and is detrimental for MBE processes.

XPS studies show that a thin oxide layer is formed (of the order of several nanometers) after chemical treatment of the samples. Carbon content data show that both the preliminary preparation of the substrates and the quality of the protective oxide layer are very important.

References

CASEY, H. C., PANISH, M. B.: Heterostructure Laser, London 1978
MOLODJAKOVA, L. P., ANDREEV, V. M.: Неорганические материалы **13**, (7) (1977)

(Received July 15, 1987)

Author's address:

c/o Dr. L. D. PRAMATAROVA
Institute of Solid State Physics. Bulgarian
Academy of Sciences
Boulv. Lenin 72
1784 Sofia, Bulgaria

Cryst. Res. Technol.	23	1988	1	K16–K19

D. K. GHOSH, L. K. SAMANTA, G. C. BHAR

Burdwan University, Deptt of Physics, Laser Laboratory, Burdwan, India

Optical Properties of Some Quaternary Copper Chalcopyrites

The ternary chalcopyrite semiconductors show considerable technological potential for a wide variety of opto-electronic device applications (SHAY, WERNICK) but their application is limited only to a very few crystals because of lack of availability in suitable size and quality due to the complexity in crystal growing process. The quaternary mixed chalcopyrites offer the advantage in tailoring suitable values of refractive index, nonlinearity, lowest energy band gap and lattice parameters via the proper choice of the composition parameter. Whereas the Ag-chalcopyrites as well as their mixed versions are useful for nonlinear laser devices (BHAR et al.; SAMANTA et al.) the mixed Cu chalcopyrites are not so useful in this respect for lack of birefringence. But they are promising candidates in heterostructure LED and solar cells since the electrical conductivity can be varied over a wide range.

In this note we report the results of our calculation of optical nonlinearity, lowest energy band gap, and refractive indices of the following important Cu-mixed compounds viz. $CuGa(Se_{1-x}Te_x)_2$, $CuIn(Se_{1-x}Te_x)_2$, $CuGa_{1-x}In_xSe_2$, and $CuGa_{1-x}In_xTe_2$. Usually a material with high refractive index has a higher nonlinearity and this can be set to a desired value by the proper choice of cationic and or anionic variations. Out of the several approaches the bond change model of LEVINE has been applied successfully to a wide variety of crystals and SAMANTA et al. have extended the same with suitable modification evaluated the nonlinearity of some binary mixed crystals. In this context we also have evaluated spin-orbit splitting (Δ_{so}) and the crystal field splitting parameters (Δ_{cf}) of such mixed systems which determine/influence many other physical properties of the crystals.

The expression for Miller's delta (Δ_{ijk}) for the four bonds of the quaternary compound can be written as

$$\Delta_{ijk}(x) = F[\Delta_{ijk}^{A-C}[C(x)] + \Delta_{ijk}^{A-C}[E_h(x)] + \Delta_{ijk}^{B-C}[C(x)] + \Delta_{ijk}^{B-C}[E_h(x)]] \quad (1)$$

where

$$F\,\Delta[C(x)] = \frac{17.51 e^{-K_s(x)\, r_0(x)} [\chi_b(x)]^2\, C(x)\, V_b(x)}{E_g^2(x)\, d^2(x)\, [q(x)/e][\chi_{compd}(x)]^3}$$

and

$$F\,\Delta[E_h(x)] = \frac{1.17 f_c(x)\, [\chi_b(x)]^2\, \varphi(x)\, V_b(x)}{d(x)\, [q(x)/e]\, [\chi_{compd}(x)]^3}$$

with the macroscopic susceptibility for typical mixed system $AB[(C_1)_x\,(C_2)_{1-x}]_2$ given by

$$\chi_{compd}(x) = \frac{X}{2}[\chi^{A-C_1} + \chi^{B-C_1}] + \frac{1-x}{2}[\chi^{A-C_2} + \chi^{B-C_2}] \quad (2)$$

and the mean radius of the bond and the mean bond radius for the type $A - [(C_1)_x \cdot \cdot (C_2)_{1-x}]$ given by

$$r_0(x) = \frac{1}{2} [r_A + x r_{C_1} + (1 - x) r_{C_2}] \tag{3}$$

All other quantities in the expression have their usual meaning as has been defined by Levine and properly modified by Samanta et al. for applying to the mixed crystals, and the method of evaluation of different parameters has been discussed in the works of the latter. The quantity Δ_{so} has been found out following the method of Hubner and Unger after modifying properly for the mixed system; while Δ_{cf} by introducing the effect of d-electrons and the lattice parameters a, c from Avon et al. for the mixed system. Now the values of Δ_{so} and Δ_{cf} together with the expression (3) of Samanta et al. help one to evaluate the lowest energy band gap of the mixed system with concentration, while the refractive index at different concentration can now be found out following the relation suggested by Ghosh et al. given by

$$n^2 - 1 = \frac{A}{(E_g + B)^2} \tag{4}$$

where A is the band gap dependent parameter and B a band gap additive constant remaining unchanged for a particular group of material, the details of which has been shown in Table 1 of Ghosh et al.

The results of our calculation of optical nonlinearity, spin-orbit splitting, crystal field splitting energy, refractive index, and band gap variation are shown in Table 1.

Table 1
Calculated values of nonlinearity, spin Orbit splitting, crystal field splitting, refractive index, and lowest band gap, together with the available experimental values

Compound	concentration x	nonlinearity $\Delta X10^6$ e.s.u.	spin-orbit splitting energy Δ_{so} (eV)	crystal field splitting energy Δ_{cf} (eV)	refractive index n	lowest energy gap E_g (eV)	
						calculated	expt.
$CuGa(Se_{1-x}Te_x)_2$	0	0.75	0.28	0.11	2.71	1.66	1.66[a])
	0.25	0.85	0.40	0.09	2.72	1.55	
	0.50	0.96	0.52	0.07	2.74	1.45	
	0.625	1.01	0.58	0.06	2.75	1.40	1.44[a])
	0.75	1.08	0.64	0.06	2.76	1.33	
	1.00	1.19	0.75	0.05	2.77	1.22	1.24[a])
$CuIn(Se_{1-x}Te_x)_2$	0	0.82	0.35	0.01	2.77	0.98	0.98[a])
	0.17	0.89	0.44	0.01	2.77	0.98	
	0.25	0.97	0.48	0.00	2.77	0.99	
	0.50	1.13	0.60	0.00	2.76	1.00	
	0.75	1.30	0.72	0.00	2.76	1.01	1.05[a])
	1.00	1.46	0.84	−0.01	2.76	1.02	1.05[a])
$CuGe_{1-x}In_xSe_2$	0	0.75	0.28	0.11	2.62	1.66	1.66[b])
	0.25	0.75	0.30	0.06	2.65	1.50	1.46[b])
	0.50	0.76	0.32	0.03	2.69	1.33	1.29[b])
	0.75	0.78	0.34	0	2.73	1.16	1.13[b])
	1.00	0.82	0.35	0	2.76	1.02	0.99[b])
$CuGa_{1-x}In_xTe_2$	0	1.19	0.75	0.05	2.71	1.22	1.24[c])
	0.25	1.23	0.77	0.02	2.72	1.17	
	0.50	1.29	0.80	0.00	2.74	1.11	
	0.75	1.38	0.82	−0.005	2.76	1.04	
	1.00	1.49	0.84	−0.01	2.77	0.98	0.98

[a]) Loferski et al. (1981), [b]) Bodner et al. (1982), [c]) Shay and Wernick (1975)

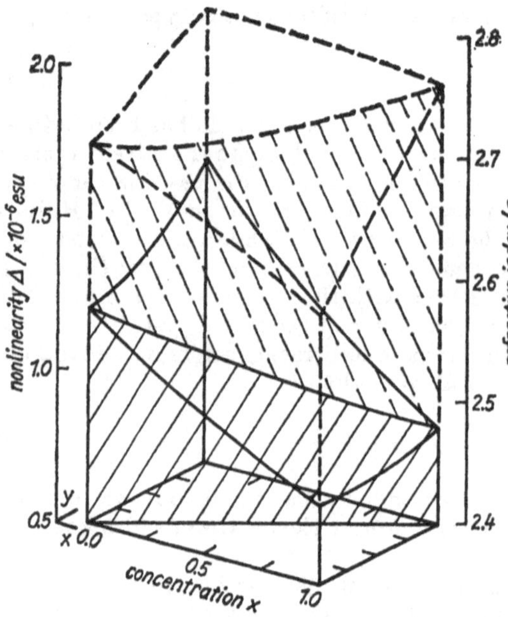

Fig. 1. Plot of the calculated nonlinearity and refractive index vs composition of the quaternary and quinary copper chalcopyrites. Full curve – nonlinearity, dashed curve – refractive index. The area shown by full lines represents the nonlinearity and the shaded, the refractive index of $CuGa_{1-x}In_xSe_xTe_{1-x})$ system

The calculated values of band gap agree well with the experimental values. As regards the nonlinearity, it is seen that for a fixed anion, it increases as Ga is replaced by In, also for the compound with both the anion fixed, the former trend is also repeated. Whereas for a particular mixed system, the nonlinearity increases as anions with lower atomic weight is replaced by the corresponding higher one. The calculated nonlinearity for all the mixed system studied is found to obey a parabolic variation with concentration. For the mixed quinary $CuGa_{1-y}In_y(Se_xTe_{1-x})_2$ the calculated nonlinearity for $y = x$ form can be represented by a power series formula.

$$\Delta = 1.19 - 0.42x + 0.09x^2 - 0.05x^3 \tag{3}$$

A three dimensional plot of the calculated nonlinearity and refractive index of the above mixed compounds including the quinary $(y - x)$ have been shown in Figure 1.

Considering the band gap variation with concentration it appears that amongest the mixed compounds studied, the $CuGa(Se_{1-x}Te_x)_2$ has got the desired value of the lowest energy band gap ~ 1.4 eV corresponding to the concentration $x = 0.625$ which agrees well with the value suggested by Romeo for photo-voltaic energy conversion for the same. It is interesting to note in this connection that the calculated values of lowest energy band gap agree well with the experimental values for other mixed system also.

Thus it is found that the more detailed study about the band gap, refractive index band structure parameters would be very much helpful in selecting for appropriate materials not only in optoelectronic devices but also in photo-voltaic solar cells.

References

Avon, J. E., Yodee, K., Wolley, J. C.: J. Appl. Physics **55** (1987) 524
Bhar, G. C., Ghosh, D. K., Ghosh, P. S., Schmitt, D.: Appl. Opt. **22** (1983) 2492
Bodner, I. V., Bologa, A. P., Korzum, B. V.: phys. stat. sol. (b) **109** (1982) K31
Ghosh, D. K., Samanta, L. K., Bhar, G. C.: Infrared Phys. **24** (1984) 43

HÜBNER, K., UNGER, K.: phys. stat. sol. (a) B **54** (1972) K65
LEVINE, B. F.: J. Chem. Phys. **59** (1973) 1463, Phys. Rev. B **7** (1973) 2591
LOFERSKI, J. J., KUIETNISK, M., PIEKOAZEWJKI, J., SPITZER, M., ARYA, R., ROESALAR, B., BEAULIER, R., VERA, E., SCHWOHUM, J., KAZMERSKI, L. L.: Proc. 15th IEEE photo voltaic specialists conf. Orlando, Flouride, New York 1981
ROMEO, N.: Jap. J. Appl. Physics **19** Suppl. (1980) 19-3, 5
SAMANTA, L. K., GHOSH, D. K., BHAR, G. C.: Phys. Rev. B **33** (1986) 4145
SAMANTA, L. K., GHOSH, D. K., BHAR, G. C.: phys. stat. sol. (a) **93** (1986) K51
SHAY, J. L., WERNICK, J. H.: Ternary Chalcopyrite Semiconductors: Growth, Electronic properties and Applications, New York 1975

(Received July 17, 1987)

Authors' address:

c/o Dr. G. C. BHAR
Burdwan University, Department of Physics
Laser Laboratory
Burdwan 713104, India

Cryst. Res. Technol.	23	1988	1	K20–K24

J. AULEYTNER, J. BAK-MISIUK, Z. FURMANIK, J. MORAWIEC

Institute of Physics, Polish Academy of Sciences, Warsaw, Poland

On the Possibility of Application of X-ray Diffraction Edge Contrast for the Quantitative Determination of High-energy Heavy Ion Range in Silicon

1. Introduction

The purpose of this work was to estimate the distribution and the range of structural changes caused by Kr ions with 42 MeV/n energy in silicon basing on the quantitative measurements of diffraction edge contrast in dependence of the ion path in crystal.

In the previous work the authors showed that the diffraction edge contrast is very sensitive on the structural changes in single crystals caused by high energy heavy ions, even in the case of small dose (AULEYTNER et al.).

For studies the X-ray transmission topography method was applied to obtain contrast image caused by stresses between implanted and nonimplanted regions. The relative intensities of contrast were photometric assigned, measuring film darkning which is quantitatively described as the blacking. To determine the character of structural changes the Bond method was used. It let us to estimate the value of stresses responsible for creating of diffraction contrast in the plate with highest defect concentrations.

2. Experimental and results

In our experiment the six parallel-cut silicon plates of the thickness of about 190 μm were used. All the plates were rectangles with edges parallel to ⟨110⟩ and ⟨112⟩ directions, respectively. Both sides of the plates were chemically polished to remove the damaged surface layer which could be produced during mechanical treatment. It was proved by X-ray transmission topography that before implantation the plates were dislocation- and stressfree. The plates were placed one behind another in the copper holder in such a way, that the ion beam was parallel to the ⟨111⟩ direction. The samples were covered by a 2 mm thick copper plate with a circular hole of 8 mm diameter (Fig. 1). During the irradiation the holder was cooled to 77 K with liquid nitrogen vapour. The beam of krypton ions of 42 MeV/n energy from heavy ion accelerator "GANIL" (Caen, France) was used. The irradiation dose was 3×10^{13} i/cm². The implanted plates were studied with the Lang type X-ray transmission topography. The $K\alpha_1$Mo radiation emitted from X-ray tube with quasi point source was used. All topogrammes were obtained in the same time by the voltage of 40 kV and anode current of 300 μA.

Figure 2 shows the diffraction image of the fourth plate. The strong edge contrast is visible and also differentiated contrast in the whole imaged region caused by the stresses and some "surface" defects.

Figure 3 shows the parts of topogrammes from which the photometric curves of the blacking at the same level (A–B line) were obtained. The arrow indicates the direc-

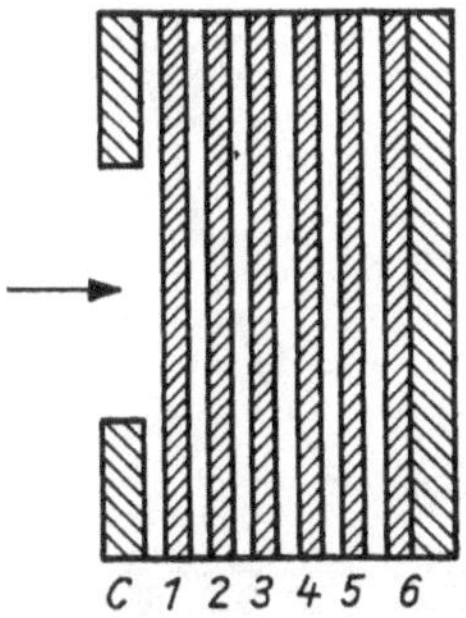

Fig. 1. Arrangement of Si plates (1 – 6) during implantation. C – copper plate

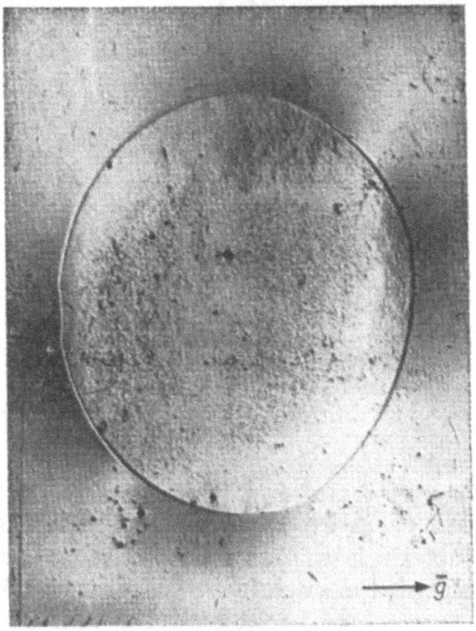

Fig. 2. Topogramme of plate number 4

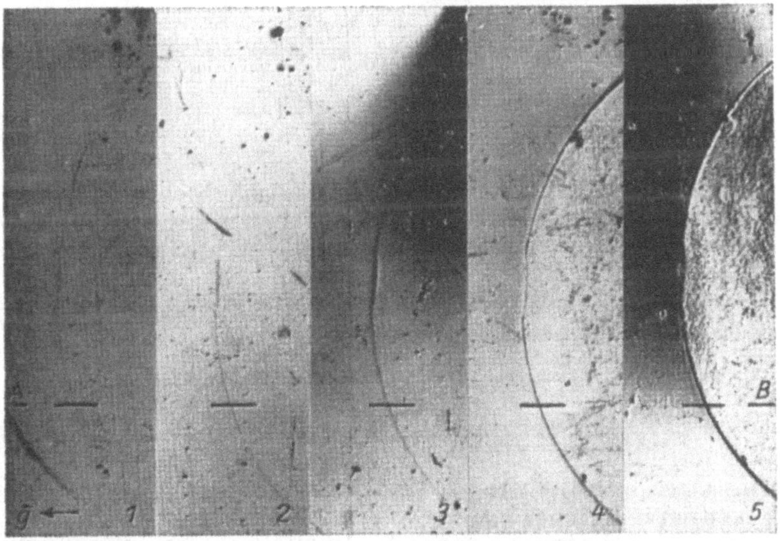

Fig. 3. The parts of topogrammes of plates (1 – 5). Reflection 422, Mo$K\alpha_1$ radiation, g – direction of diffraction vector; A – B the level on which the photometric measurements of contrast intensity were performed

tion of diffraction vector (in this case ⟨211⟩). There is visible the increase of contrast intensity as ion beams pass through larger and larger numbers of plates. No edge contrast for plate 6 was observed. Determination of area under photometric curves after removing of background let to estimate the magnitude of relative integrated intensities of contrast satysfaying the condition when intensity is proportional to the blacking in discussed range. Figure 4 shows the dependence of the integrated intensity of contrast on the path length of the ions in the material. The curve I corresponds to this side of crystal "output one", which was turned towards the film (during taking the

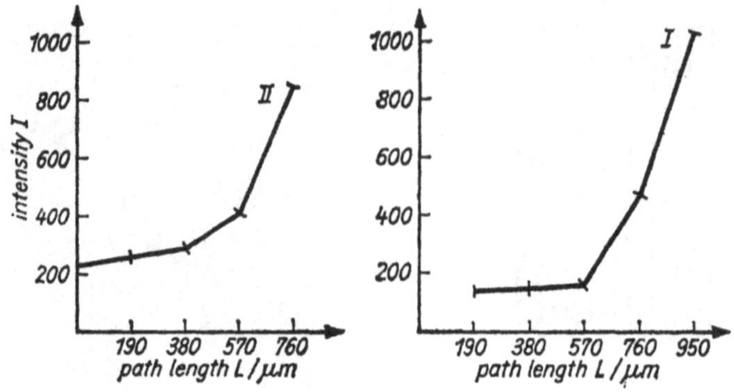

Fig. 4. (I, II) Dependence of the integrated intensity I of contrast on the path length (L µm) of Kr ions

topogramme) and, the curve II corresponds to the side bombarded by ions which was turned towards film. Very similar courses of obtained curves show that applied method is enough sensitive for measuring the defect distribution. In the fifth plate the maximum values of contrast intensities were observed. It testifies that just in this plate most of ions stopped their running, loosing whole energy. In the case of this plate it succeded to measure using the Bond method the increase of lattice constant in the ⟨111⟩ direction about: 0.000196 Å what corresponds to the change of lattice constant in ⟨112⟩ direction: $0.18 \cdot \Delta d_{111} = 0.0000352$ Å (WORTMAN, EVANS).

3. Discussion

The change of sample volume measured in implanted region is caused by lattice defects. Basing on the carried out measurements we can ascertain that maximum range of ions, in that case, is in the interval of 760–950 µm, whereas the theoretical calculated range is 716 µm.

Knowing the deformation, Young's module and Poisson's ratio we can estimate the stresses connected with the observed contrast.

$$\varepsilon_p' = \sum_q S_{pq'}\sigma_q' \; (p, q\ 1, 2, \dots 6\ .)$$ (WORTMAN, EVANS)

where σ_q' and ε_p' determine the relative stresses and strains, S_{pq} – compliance coefficients.

In discussed case the stresses are 2.96×10^7 dyn/cm^2 (sample 5). In the plates 1–4 the stresses are much more smaller.

Increase of sample volume during implantation process by ions with small dose is mainly caused by point defects particularly their clusters. This was shown by using the transmission electron microscope. Figures 5a and b show the image of crystal interior (plate 4) as a bright field. Defects with densities in the range from 9×10^9 to 9×10^{10} cm^{-2} are visible. The diameter of defect is about 100 Å. Figure 5a shows the image in the case overfocus (dark spots), in Figure 5b underfocus – bright spots. That kind of contrast can be obtained in the case of agglomeration of vacances. Studying the characteristic X-ray radiation of krypton no signal was observed. That means, that vacance agglomerations are either empty or gas content is very small. The similar changes were observed in the case of metal irradiation (QUERÉ).

Summarizing, we have shown that measurements of contrast intensity in the X-ray image of the boundary between implanted and nonimplanted parts of crystal let to

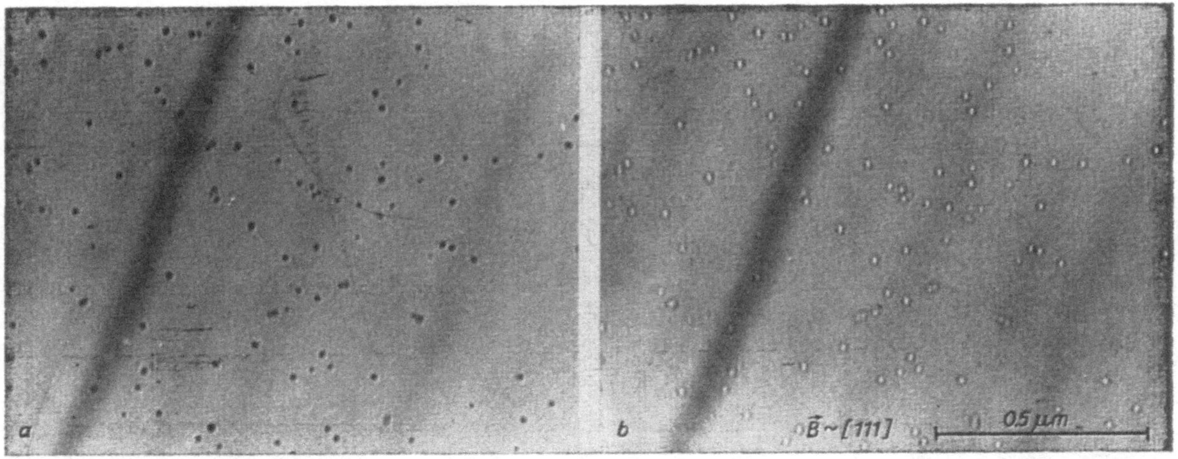

Fig. 5. Overfocused a) underfocused, b) many beam images of spherically shaped defects – vacances agglomerations or voids in silicon target implanted with Kr ions, the same areas, (TEM-JEM 200 A operating at 200 kV)

estimate approximately the profile of depth distribution of defects created during implantation process.

In discussed case of Kr ions the estimated maximum range is larger than theoretical calculated one. Using the Bond method it was stated that in the range corresponding to the maximum range the increase of lattice contrast of $\Delta a_{111} = 0.000196$ Å was observed. In this case the calculated stress parallel to the diffraction vector ⟨112⟩ was 2.96×10^7 dyn/cm². Defect images were obtained using TME on the depth of 570 to 760 μm. These defects were interpreted as an agglomeration of vacances or voids with diameter of ca 100 Å.

The authors would like to thank Dr. J. C. Jousset and Dr. M. Toulemonde for carrying out the implantation, Prof. Y. Queré for his interest and for stimulating discussions and Msc. J. Górecka for sample preparation.

This work has been partially financed by the Institute of Experimental Physics of Warsaw University in the programme CPBP 0106.

References

AULEYTNER, J., BAK-MISIUK, J., FURMANIK, Z.: Cryst. Res. Technol. **22** (1987) K59

WORTMAN, J., EVANS, R. A.: J. Appl. Physics **36** (1) (1965) 153

QUERÉ, Y.: Proceedings of IV International Summer School on Defects, Zakopane, 18—30 June, 1973, p. 11

(Received August 3, 1987)

Authors' address:

c/o Prof. Dr. J. AULEYTNER
Institute of Physics, Polish Academy of Sciences
Al. Lotników 32/46
PL-02 668 Warsaw, Poland

W. SCHMITZ, J. KORNATOWSKI, G. FINGER

Wissenschaftsbereich Kristallographie, Sektion Chemie der Karl-Marx-Universität Leipzig, and Institute of Physics, Nicolaus Copernicus University Toruń, Poland, and Zentralinstitut für Physikalische Chemie der Akademie der Wissenschaften der DDR Berlin

Growth of NaX Zeolites in the Presence of Triethanolamine (TEA)

In honour of Professor Hermann NEELS 75th birthday

1. Introduction

A method for preparing larger crystals of the synthetic zeolites NaA and NaX has been reported by CHARNELL. Basing on this procedure, we investigated detailed factors which influence the crystallization process considering chemical and crystallographic approachs to the problem (GUTSZE et al.; KORNATOWSKI et al. (NaA); SCHMITZ et al. (NaA)). Synthesis of NaX is more complicated than that for type A because the crystallization of other zeolite phases is much harder to avoid. The number of factors operating in the NaX system is larger and renders more difficult to assign to nucleation or crystal growth the experimental parameters.

The aim of this work was to study the growth of NaX zeolite under the influence of different experimental conditions as it had been done for NaA KORNATOWSKI et al. and SCHMITZ et al. In this paper we want to present our results dealing especially with the crystal growth of NaX zeolite. A more extensive report on chemical aspects containing plan and tables of results has been given by KORNATOWSKI et al. (NaX).

2. Experimental

All experimental conditions except the gel compositions were exactly the same like for type A (KORNATOWSKI et al. (NaA)).

The conditions were changed in the series of experiments in comparison with the CHARNELL's data.

The main points of our investigations concerned:

- different starting materials and mixing conditions
- higher dilution (25%)
- systematic decrease of aluminate content
- systematic increase of TEA content
- preparation of double samples.

3. Results

An interesting fact results from preparation of double samples. The reproducibility of the results, i.e. yield of NaX, yield of Ba and A impuring zeolites, and crystal sizes of NaX became the worse the more different were the gel concentrations from the concentration used by CHARNELL.

Other results are:

Higher dilution has no marked influence on crystal size. It causes low decrease of impuring phases content.

Decrease of sodium aluminate content also does not influence crystal size of NaX but the yield of NaX decreases in favour of B zeolite. Accompaning effect is the change in the Si/Al ratio from 1.1 to 1.6 (cf. KORNATOWSKI et al. (NaX)). A higher ratio could not be observed. The gels, containing the lowest concentrations of aluminate, crystallize almost only B zeolite.

Additional TEA amount (50–250%) does not markedly influence both the crystal size and the Si/Al ratio. The best yields of large crystals were obtained when additional amounts of TEA were low. More than 100% additional TEA led to loss of NaX crystallization.

In opposite to our results concerning NaA crystallization, it was found that NaX growing system is most sensitiv for starting materials used as well for the methods of the gel formation and treatment. We examined the influence of different starting materials and methods of mixing of the initial solutions, ultrafiltration of the solutions, time of mixing and of shaking the gel during the crystallization period following the concentrations given by CHARNELL (cf. KORNATOWSKI et al. (NaX)). These factors are essential for the success in growing larger NaX crystals. The best results were obtained (NaX crystals up to 135 μm) when silica sol and $Al(OH)_3$ were used as starting materials and the aluminate solution was dropped into the silicate solution at constant stirring. Other materials and methods of mixing worked worse in the NaX yield and crystal size. Ageing of the gel should not be employed under the conditions used by us.

4. Discussion

The main result of our investigations on NaX zeolite crystallization is the statement that it is nearly impossible to improve yield and crystal size by altering the experimental parameters which have a marked influence on the NaA crystallization. For that, it could be stated that the nucleation was not completely finished when the gel had been formed under the influence of TEA (SCHMITZ et al.).

Concerning the NaX crystallization the process is different. FREUND investigated the NaX crystallization without TEA. He pointed out that secondary nucleation in the NaX forming system is only possible by collision breeding of a stagnant layer associated with each nucleus. This can be provoked by stirring over the full crystallization period not used in our experiments because TEA causes prolongation of the period from some hours to many weeks. Shaking the samples during the crystallization period led to the secondary nucleation in some cases as indicated by two maxima of the sizes distributions in some of our samples.

Thus, it is to consider the same mechanism of the crystallization of the zeolite NaX when TEA is used. The results confirm that the crystallization products are strongly determined by initial period of NaX nucleation during the gel formation. Only the experimental parameters which touch the nucleation of NaX exert an influence on the yield and size of NaX. These parameters are starting materials and methods of the solutions mixing, and much less concentrations of water, sodium aluminate or TEA. The nucleation process is more difficult to controll than the crystal growth as shown by different reproducibility obtained by preparation of double samples.

NaX nucleation takes place best at low temperature whereas the crystal growth is accelerated at higher temperature (70 – 90 °C) and, in addition formation of zeolite B is preferred (FREUND). The number of NaX nuclei must be sufficient that their growth can prevent spontaneous nucleation of B zeolite to such an extent that B zeolite do not dominate the overall reaction. On the other hand to many nuclei lead to smaller crystals. The art to grow larger NaX crystals is to keep the balance between these two opposite working aspects.

Fig. 1. NaX zeolite octahedra

In the case of the NaA crystal growth, we have been able to influence the growth rate by changing the gel composition (KORNATOWSKI et al.; SCHMITZ et al.). The lowering of the NaX growth rate appears to be not easily influenced.

The driving force for the growth of zeolite crystals is the higher solubility of the solid gel particles in the solution covering the gel. The difference in the solubilities of the amorphous and crystalline phase is lower for NaX than for NaA. This is evident from the occurrence of rounded crystal forms of NaX (Fig. 1) resulting from dissolving process never observed on NaA crystals. As a consequence of low growth rate, NaX can grow larger than NaA crystals first obtained by CHARNELL when attention is payed to the nucleation process.

From our experiments we may recommend for successfull growth of larger NaX crystals under the consideration of different reproducibility:

1. Following the concentration given by CHARNELL.
2. Testing and using the best available starting materials.
3. Dropping the aluminate solution into the silicate solution.
4. Use of ultrafiltration.

This work was partially supported by the Polish Ministry of Science and Higher Education, Project CPBB 01.06.

References

CHARNELL, J. F.: J. Cryst. Growth 8 (1971) 291

FREUND, E. F.: J. Cryst. Growth 34 (1976) 11

GUTSZE, A., KORNATOWSKI, J., NEELS, H., SCHMITZ, W., FINGER, G.: Cryst. Res. Technol. 20 (1985) 151

KORNATOWSKI, J., FINGER, G., SCHMITZ, W.: Pol. J. Chem. (in press) (NaA)

KORNATOWSKI, J., FINGER, G., SCHMITZ, W.: Pol. J. Chem. (in press) (NaX)
SCHMITZ, W., KORNATOWSKI, J., FINGER, G.: Cryst. Res. Technol. **22** (1987) 35

(Received August 26, 1987)

Authors' addresses:

Dr. W. SCHMITZ
Wissenschaftsbereich Kristallographie, Sektion Chemie der Karl-Marx-Universität Leipzig
Scharnhorststr. 20
Leipzig
DDR-7030

Dr. J. KORNATOWSKI
Instytut Fizyki UMK
ul. Grudziadzka 5
87-100 Toruń, Poland

Dr. G. FINGER
Zentralinstitut für Physikalische Chemie der Akademie der Wissenschaften der DDR
Rudower Chaussee 5
Berlin-Adlershof
DDR-1199

Continued from cover page 4

Short Notes

Crystal Research and Technology
Volume 23 1988 Number 1

Contents

Review

Original Papers

Continued on cover page 3

Crystal Research and Technology is indexed in Current Contents/PC & ES.

MIX
Papier | Fördert gute Waldnutzung
FSC® C083411

Zeitfracht Medien GmbH
Ferdinand-Jühlke-Straße 7
99095 Erfurt, Deutschland
produktsicherheit@kolibri360.de